JMYK

INSTITUTE OF PSYCHIATRY
Maudsley Monographs

MAUDSLEY MONOGRAPHS

HENRY MAUDSLEY, from whom the series of monographs take its name, was the founder of the Maudsley Hospital and the most prominent English psychiatrist of his generation. The Maudsley Hospital was united with the Bethlem Royal Hospital in 1948, and its medical school, renamed the Institute of Psychiatry at the same time, became a constituent part of the British Postgraduate Medical Federation. It is entrusted by the University of London with the duty to advance psychiatry by teaching and research.

The monograph series reports work carried out in the Institute and in the associated Hospital. Some of the monographs are directly concerned with clinical problems; others, less obviously relevant, are in scientific fields that are cultivated for the furtherance of psychiatry.

Joint Editors

Professor Gerald Russell
MD, FRCP, FRCP (Ed),
FRC Psych.

Professor Edward Marley
MA, MD, DSc., FRCP,
FRC Psych.

Assistant Editor

Dr Paul Williams
MD, MRC Psych., DPM

with the assistance of

Miss S. E. Hague, BSc. (Econ), MA

INSTITUTE OF PSYCHIATRY

Maudsley Monographs

Number Thirty-Two

Sentenced to Hospital
Offenders in Broadmoor

By

SUSANNE DELL, MA

*Lecturer, Forensic Psychiatry Section, Department of Psychiatry,
Institute of Psychiatry, London SE5 8AF*

GRAHAM ROBERTSON, PhD

*Lecturer, Forensic Psychiatry Section, Department of Psychiatry,
Institute of Psychiatry, London SE5 8AF*

OXFORD UNIVERSITY PRESS

1988

Oxford University Press, Walton Street, Oxford OX2 6DP

Oxford   New York   Toronto
Delhi   Bombay   Calcutta   Madras   Karachi
Petaling Jaya   Singapore   Hong Kong   Tokyo
Nairobi   Dar es Salaam   Cape Town
Melbourne   Auckland
and associated companies in
Berlin   Ibadan

Oxford is a trade mark of Oxford University Press

Published in the United States
by Oxford University Press, New York

© Institute of Psychiatry 1988

British Library Cataloguing in Publication Data
Dell, Susanne
Sentenced to hospital: offenders in
Broadmoor.—(Maudsley monographs, 32).
1. Berkshire Crowthorne. Special hospitals.
Broadmoor Special Hospital. Patients
I. Title   II. Robertson, Graham
III. University of London. Institute of
Psychiatry   IV. Series
362.2 ´ 1
ISBN 0–19–712156–X

Library of Congress Cataloging in Publication Data
Dell, Susanne.
Sentenced to hospital: offenders in Broadmoor/by Susanne Dell,
Graham Robertson.
(Maudsley monographs; no. 32.)
Bibliography: p.
Includes index.
1. Broadmoor Hospital (Crowthorne, Berkshire)   2. Insane, Criminal
and dangerous—Great Britain.   I. Robertson, Graham, 1946–
II. Title.   III. Series.
HV8742.G73B73   1988   365´.942294—dc19   88–4025
ISBN 0–19–712156–X

Set by Downdell Ltd., Abingdon, Oxon.
Printed in Great Britain
at the University Printing House, Oxford
by David Stanford
Printer to the University

Acknowledgements

The grant for this work came from the Department of Health and Social Security, and without it the study could not have been undertaken. The award was to the Forensic Unit at the Institute of Psychiatry and, in particular, to Dr John Hamilton (then a full time senior lecturer in that unit) and to Professor John Gunn, its head. It is they who were responsible for our employment and for the proposal to examine the treatment of psychopaths in Broadmoor. We would like to express our gratitude to them but should make it clear that responsibility for the contents of this book is ours alone.

Shortly after the study began, Dr Hamilton was appointed Director of Broadmoor Hospital and, along with his colleagues there, had the burden of being interviewed by us. As each man in the study required a separate interview, this was no easy task. Despite the heavy work load which they already bear, the consultants cooperated fully in the project though there must have been times when they could have wished us elsewhere. It is difficult to find words with which to express our gratitude to them. It is to their frankness and willingness to discuss treatment and detention issues that we owe much of our understanding of these subjects. That the doctors were prepared to share their experience and to talk bluntly about what they thought was wrong as well as what was good about Broadmoor meant that we became aware of many issues that we might otherwise have overlooked. We are deeply grateful to them.

Much of the burden for arranging the interviews with the consultants fell on their secretaries and despite their many other duties they accommodated us with every courtesy. The same cooperation was given by everyone in the hospital, a fact which made the study a pleasure to undertake. The nursing staff at all levels were not only willing to arrange appointments for us to interview men but also to discuss any issues which we wanted to raise; their generosity of time is greatly appreciated.

Most departments in the hospital helped us from time to time. Particular mention must be made of three of these: the Medical Records Office, the Psychology Department, and what we have termed the Jenny France department.

From our first day in Broadmoor we were dependent on Medical Records for guidance in finding notes and for allowing us use of their office for innumerable searches. The Psychology Department provided free and full access to their records, again something which we have not taken for granted. Their friendliness and support made an otherwise tedious task thoroughly enjoyable. Finally, the contribution of Mrs Jenny France to the project must be acknowledged. Mrs France allowed us to share her office during the two years it took to complete our work and was, on countless occasions, of assistance in providing information despite her extremely heavy work load. To all the staff of Broadmoor then we extend our thanks, and that gratitude is owed in equal measure to the men we interviewed. Although all patients were assured of the confidential nature of the study, they had nevertheless to take that matter on trust and we were constantly surprised at how willing they were to have their privacy intruded upon by complete strangers.

We are most grateful to Dr Paul Bowden and Dr Adrian Grounds for commenting on earlier versions of our report. We wish also to thank Mrs Elizabeth Parker and her staff for providing access to the data in the Special Hospital Case Register. The editors of the Monograph series suggested many useful changes. Maureen Bartholomew and Mollie Loshack shared the onerous task of typing the manuscript and to them we are, as always, very indebted.

June 1987   S.D.
London   G.R.

Contents

Tables

1 Introduction

BROADMOOR

The Criminal Lunatic asylum at Broadmoor was opened in 1863, the country's first state institution for mentally abnormal offenders. The events which led to its establishment have been traced by Walker and McCabe (1973): chief among them was concern that mentally disordered persons involved in serious violence should be appropriately and humanely treated in conditions of adequate security.

The hospital was built some 35 miles from central London on the summit of a Berkshire ridge, in accordance with the then fashionable belief in the therapeutic powers of country air. It catered for three types of patients: those who were in custody as a result of being acquitted by reason of insanity, those who were found insane on arraignment, and those who were found insane while serving a penal sentence. The hospital was owned by the Home Secretary and managed on his behalf by a council of supervision.

Broadmoor's history until 1952 has been chronicled by Partridge (1953) and will not be repeated here. Major and enduring changes to the legal framework in which the hospital operated were introduced by the 1959 Mental Health Act. This designated Broadmoor as one of three special hospitals (Rampton and Moss Side being the others), put them all under the direct management and control of the Ministry of Health, now the Department of Health & Social Security (DHSS), and made them available for patients subject to detention under the Mental Health Act, who in the opinion of the Minister required treatment under conditions of special security because of their dangerous, violent or criminal propensities (Mental Health Act 1959 Sections 97–99). Broadmoor was thus for the first time enabled to take patients who had not been through the courts. In addition, it began to receive patients under the new hospital order provisions of the 1959 Act. These enabled the courts to send to hospital offenders who had committed imprisonable offences, if they were suffering from mental illness, psychopathic disorder, or mental handicap. Once hospitalized, the discharge of these offender patients, unless ordered by a Mental Health Review Tribunal, became a matter for the hospital authorities. However, if the offender had been tried in the higher courts, and the court thought it necessary for the protection of the public, the Act enabled a restriction order to be attached to the hospital order. The effect of this was to prevent the hospital (or a tribunal) from moving the patient without the consent of the Home Secretary. The restriction order also gave the Home Secretary power to impose conditions of discharge, and to recall released patients.

Once these hospital order provisions were enacted, it was not long before the majority of Broadmoor's admissions were made under them, and offenders subject to hospital orders with restrictions have for many years now accounted for most of Broadmoor's intake. The number and type of male admissions to Broadmoor in recent years is shown in Table 1.1.

The 1983 Mental Health Act made no changes that were specific to the special hospitals, and it preserved in broad outline the hospital order provisions of the previous Act. It did however introduce a number of new safeguards for detained patients: one of these was to give to Mental Health Review Tribunals the power in specified circumstances (section 73) to discharge restricted offenders.

It has been pointed out that there is often confusion about the role of the Home Office in relation to Broadmoor. The present Medical Director reminds us (Hamilton, 1985) that 'Special hospitals are not prison hospitals; there are no prison officers and all patients are detained under mental health legislation'. However, the Home Office influence in Broadmoor remains important. The great majority (79 per cent) of patients cannot leave the hospital, even for a day's outing, without Home Office consent. They are either subject to hospital orders with restrictions (62 per cent of Broadmoor's resident population); or they have been found by the courts to be either unfit to plead or not guilty on grounds of insanity (5 per cent of the population), in which case the Home Office has control of their movement under the Criminal Procedure (Insanity) Act, 1964; or (12 per cent of the population) they are transferred mentally disordered prisoners whose release cannot be effected without Home Office permission (Mental Health Act, 1983, section 49): (figures are for 1984, DHSS, 1984). Apart from the Home Office's role in controlling the release of patients, its dominant historical influence is evidenced by the fact that Broadmoor's nurses still wear the boots, hats, and uniforms characteristic of prison guards, and that the trade union to which they belong is the Prison Officers' Association.

Security is the paramount factor in Broadmoor's life, as it is the paramount factor in physical terms. With its massive boundary walls, the hospital (now being rebuilt) looks like a nineteenth-century prison, naturally enough, since it was designed by Joshua Jebb, architect of Pentonville. Today Broadmoor holds some 500 patients from all parts of the country; a fifth of them are women. Patients with mental handicap are not as a rule accepted. At the time of the study, eight consultants were in post, of whom two were half-time; in addition, there were some seven other medical staff, including training-grade psychiatrists. The nurse to patient ratio was 1 : 1 (Hamilton, 1985).

All admissions to Broadmoor are compulsory. This follows from the legal framework in which the hospital functions: the law (section 97 of the Mental Health Act, 1959, now superseded by section 4 of the National Health Service Act, 1977) specifies that special hospitals are 'for persons subject to detention'. In this respect, as in so many others, Broadmoor stands apart from ordinary psychiatric institutions: in local National Health Service hospitals, over 90 per cent of admissions are voluntary. Broadmoor is set apart too by the proportion of patients that have been hospitalized as a result of criminal proceedings. Only 9 per cent are held under civil procedures, over 90 per cent under criminal provisions (DHSS, 1984).

ADMISSION PROCEDURES

The Department alone has the power to authorize admission to special hospitals (National Health Service Act, 1977). In recent years it has been customary for the hospital consultants to see prospective patients and to advise the Department's officials on whether they should be accepted.

In its Memorandum on the Mental Health Act (DHSS, 1983) the DHSS has outlined the criteria it today uses in determining admission to the special hospitals. The memorandum says that a special hospital place can only be justified 'when the highest level of security is required and no lesser degree of security would provide a reasonable safeguard to the public' (para 266). The Department points out that it is the responsibility of the National Health Service authorities to provide comprehensive facilities for patients who do not need maximum security, and the memorandum states that 'any lack of local provision for difficult to place patients will not usually be accepted as a reason for admission to a special hospital' (para 269). In other words, the failure of local hospitals to accept or manage difficult patients will not 'usually' be a reason for making special hospital beds available.

Chiswick (1982) has calculated that no more than one convicted defendant in every 10,000 receives a special hospital disposal. He draws attention to the large role which chance plays in determining whether an offender is referred for such a disposal. The court, through defence, prosecution, or the bench itself may or may not seek psychiatric advice before the sentencing stage. If such advice is sought, the psychiatrists involved may or may not suggest a psychiatric disposal, let alone an application for a special hospital bed. In addition, our own research showed that special hospital consultants who report on applications to the DHSS vary greatly in their views about those who should be admitted.

Applications for special hospital beds come mainly from psychiatrists working in the forensic field, but some are made by local psychiatric hospitals in respect of patients that have proved difficult to manage. After Broadmoor doctors have visited the prospective patient, they discuss the case with their colleagues, before sending their recommendations to the DHSS. We were fortunate in being able to attend these weekly admission meetings during our research in 1982–4 and therefore to learn what considerations influenced the doctors in suggesting whether or not Broadmoor places should be made available. These issues are discussed later in this report, but three points may be noted here. First, that if a Broadmoor consultant says a patient needs a bed, then the officials of the DHSS usually make it available. Secondly, despite the words of the DHSS memorandum quoted above, special hospital consultants inevitably have regard to what would happen to a prospective patient if they did not recommend admission. If the patient is seriously ill and a National Health Service bed cannot be found, the consultants are liable to recommend a Broadmoor place, even if the security needs in the case fall short of the official guidelines.

Thirdly, it has to be remembered that the Department's admission practices have changed over the years. The Mental Health Act 1959 introduced provisions allowing the detention of patients suffering from 'psychopathic disorder' (PD), a condition which was defined (section 4(4)) as 'a persistent disorder or disability of mind . . . which results in abnormally aggressive or seriously irresponsible conduct . . . , and requires or is susceptible to medical treatment'. In the first ten or fifteen years of the Act's operation, it was not usual for DHSS officials, when bed applications were made for such patients, to ask for evidence that the psychopathic disorder was 'susceptible to medical treatment'; it was enough if the patients were described as 'requiring' it (Dell, 1984). In more recent times however, as MacCulloch (1982) has explained, the DHSS took to interpreting the phrase in the 1959 Act 'requires or is susceptible to treatment' as meaning 'requires *and* is susceptible'. Other changes have also taken place, notably in the emphasis nowadays laid on the patient's need for maximum security. There are thus resident in the special hospitals today many patients who would not be accepted under the current admission criteria.

As already noted, Broadmoor differs from ordinary psychiatric hospitals in that all its patients are compulsorily detained, and nearly all of them are held under criminal procedures. There is also another respect in which the hospital stands apart. Unlike National Health Service hospitals, Broadmoor contains substantial numbers of detained patients who are held under the legal category of psychopathic disorder. The proportion of such patients in the local hospital system is barely measurable. In Broadmoor it accounts for a quarter of the resident population. (DHSS, 1984). The detention in hospital of these patients was the focal point of our research study.

THE RESEARCH

The original purpose of the study was to examine the make up of Broadmoor's PD (psychopathically disordered) population, to look at the treatment given to these patients and the issues surrounding their discharge. Resources for the project were severely limited–two research workers (the present authors) neither of them employed full-time on the work. For this reason, it was decided to exclude female patients from the study.

In April 1982 we carried out a census of all men detained in the hospital under the legal category of Psychopathic Disorder. There were 117 such men: all were offenders, most had been admitted from the courts under hospital orders, but some were sentenced prisoners who had been transferred under the provisions of the Mental Health Act. It seemed to us that the best approach to the study of these men would be to compare them with patients who had been admitted to Broadmoor with a legal classification of mental illness. This would highlight any issues of treatment or discharge that were peculiar to the PD group, and at the same time give a more complete picture of the work that Broadmoor does. We therefore matched each PD patient on length of stay and Mental Health Act section with an offender who had been admitted as mentally ill, and these two groups of men–the psychopathically disordered and the mentally ill–were the subject of our research. All were detained under the 1959 Mental Health Act, and it was the provisions of this Act that were in force for most of the period of our research.

The work, carried out between 1982 and 1984, fell into three parts. First, information was gathered about the background of each man and of the circumstances leading to his admission. This information is presented in Chapters 2 and 6. Chapter 2 deals with the admission of mentally ill men and Chapter 6 is concerned with those classified legally as psychopathic. Both seek to give a picture of the kind of men admitted.

Our second task was to collect information about the treatment each patient had received in Broadmoor. This was done by examining his hospital case notes, by interviewing the man himself, and by cross-checking information with the therapists or department involved, where this was possible. The analysis of this information was both qualitative and quantitative. In our attempts to assess therapeutic benefit, we sought the views not only of the staff, but of the recipients of the treatment, the customers. A large part of our interviews with patients consisted of asking questions about the treatments they had received in Broadmoor and to what extent these had been found to be helpful. Chapters 3 and 7 contain the results of our inquiries about treatment, Chapter 3 being concerned with the mentally ill and Chapter 7 with the PD group.

Having traced the reasons for admission and the treatment received, we asked the consultant in charge of each man's care to indicate whether or not he considered him suitable for discharge. We then examined the various aspects of the doctors' discharge opinions. The results of this part of the study are reported in Chapters 4 and 8.

Finally, in Chapters 5 and 9, we discuss some of the themes of special interest which emerged during the course of the study. Chapter 5 is concerned with mentally ill men, and Chapter 9 with the PD group.

Our resources did not allow us to carry out any systematic observational research, and we have therefore not attempted to give any detailed account of daily life in Broadmoor. Different aspects of the regime have in recent years been described by a number of writers. (Cohen, 1981; Hamilton, 1985; Gostin, 1977). Gostin's account includes extensive extracts from the 1975 unpublished report on Broadmoor by the Hospital Advisory Service. Patients too have published their accounts of life in the hospital (Thompson, 1972; Reeve, 1983).

METHODS

We developed a number of interview schedules and questionnaires for the study. The first, entitled 'Information from Case Notes and

Medical Records' was used to record information from the hospital records. These contain information relating to the patient's admission, court reports, and Mental Health Review Tribunal proceedings as well as all statutory documentation. Case notes, kept on the ward, contained what information had been recorded about the patient since his admission, including records of ECT and medication, notes about treatments given, details of physical and mental health, and tests carried out by the psychology department. All this information was extracted for each patient.

The second questionnaire was the schedule we constructed for our interview with patients. Before we involved any patient in the study we made clear to him the purpose of our investigation. We stressed its voluntary and confidential nature. Before our meeting with each man on his ward, he had been given notice of our coming and was able at that stage, as well as during the interview, to say that he did not wish to participate. As we shall see in the relevant chapters, very few men declined to take part in the research. Most were not only willing but pleased and eager to talk about their experiences in Broadmoor, and to express their opinions about the treatment they had received.

The interview with patients lasted anything from thirty minutes to two hours. The men were seen on their own. The final questionnaire was developed from several pilot efforts to construct a schedule. Among the questions we asked were: what forms of treatment the men had received, how useful they had found them, whether they felt they had been helped by being in Broadmoor, whether they thought what had happened to them had been fair, and whether they felt ready to leave. Considerable efforts were made to establish agreement between the two interviewers at the pilot stage of the project. Many of the questions were open-ended and the construction of categories for each of these questions proved to be time-consuming; it was carried out very carefully when we were coding this information. However, agreement between us was not taken for granted, and before any analysis was undertaken a check was made as to the extent of that agreement. Despite our efforts at the beginning of the study, we found that we had rated a number of items differently, and these items were deleted from our data analyses.

The most difficult interview schedule to construct was that designed for the doctors. As with all our schedules, several pilot interviews were conducted. After a number of false starts, a schedule was constructed which we used but later abandoned. After further thought we decided to develop two schedules; one to be used if the consultant said that the man was fit to leave Broadmoor and the other, supplemented by a short item sheet completed by the doctor, if he was not considered fit for discharge. These changes resulted in some loss of data, but on balance we felt that it was preferable to sustain such loss rather than continue using a schedule with which we were not satisfied.

As was the case with the patient schedule, a number of open-ended questions were asked. A check on interviewer agreement was conducted before analysis and the procedure described in regard to the patients' interview was employed.

The main methodological issue in this study is the extensive use made of open-ended questions. The alternative strategy would have been to provide a range of responses for the interviewees, whether doctor or patient. This would certainly have saved us a great deal of work, but we eschewed this approach because we thought it unsatisfactory to limit the range of responses to the questions we were asking. Apart from the extra effort, the price to be paid for this approach was that some items turned out to be useless from the statistical standpoint. To be set against this was the fact that we gathered much richer data and that what we sacrificed in quantitative terms was more than made up for qualitatively.

Lastly, it should be stressed that our recording of treatment has been concerned with specific procedures. By this we mean the administration of drugs, ECT, speech and communication therapy, psychotherapy, whether individual or group, or one of the many treatment programmes organized by the psychology department in the hospital. We are not unmindful of the important role played by the day-to-day work of the nursing staff: to be in Broadmoor is to be in the care and control of its nurses. However, an examination of nurse–patient interaction was outside our capacity to undertake.

ORIENTATION

Finally, we would refer to the way in which this work has been presented. We became conscious during the research that the only means of doing justice to the wealth of material collected, would have been to write what would have amounted to life histories for each of the men who had taken part in the study. Lack of time, lack of space, and issues of confidentiality precluded such an exercise. However, though we have presented many tables of data related to groups and subgroups, the reader will also find many references to individual stories and circumstances. This partial flight from a purely statistical approach to the data is deliberate and is based on an awareness of the often spurious conclusions which can result when the complexity of human behaviour is reduced to manageable, but quite meaningless, numbers.

Part I: Mental illness

2 Men admitted with a legal classification of mental illness

ADMISSION MEETINGS

As explained in Chapter 1, it was usual when an application for a Broadmoor bed was made, for one of its consultants to visit the prospective patient in order to advise the DHSS on whether he should be admitted. The visits were made on a rota basis, and afterwards the cases were discussed at a weekly meeting of Broadmoor consultants. However, what recommendation was then made to the Department remained the sole responsibility of the consultant who visited.

We were fortunate in being allowed to attend the admission meetings during the period of our research. They showed us that there was considerable variation between the views of the different consultants about who should come to Broadmoor. Discussion about the potential admission of mentally ill offenders nearly always centred on the question 'What are the alternatives if we do not take him?'. Although, as we noted earlier, official DHSS admission policy seeks not to take into account the question of what alternative facilities are available, for the visiting doctors an important consideration was in fact the kind of care that prospective patients would receive if Broadmoor declined to accept them.

For example, in one of the cases discussed during the period of our research, a patient with a history of depressive illness had been remanded in custody following the attempted murder of his wife. It was not the first time he had tried to kill her. In prison, his depressive illness responded well to drugs. The local psychiatric hospital was ready to accept him. Nevertheless, the special hospital consultant who had been asked to visit thought that a Broadmoor place should be made available. His colleagues disagreed, pointing out that the patient was co-operative, had no history of absconding, and was responding to drugs. But the visiting consultant feared that as the local hospital was completely open, the patient might either get out or be allowed out and make a possibly fatal return to his wife. Admission to Broadmoor was in his view the only certain way this could be prevented, and he recommended it to the DHSS.

Another case involved a schizophrenic patient who had been convicted of assault, said not to be serious, in a local hospital. The Broadmoor consultant who visited found the man in a locked ward where conditions were quite deplorable: patients were out of control, no one seemed to be in charge, and the place was neglected and dirty. His examination of the man led him to conclude that he could be adequately managed in a well-run National Health Service ward and that therefore he should not be recommended for Broadmoor. 'We cannot accept him just because the place is so badly run. If we did, we would end up taking all the patients from all the badly run hospitals'. Most, but not all, of his colleagues agreed. One who disagreed pointed out that if the patient continued to be so badly looked after, he would deteriorate and would be likely to cause more serious harm: 'The incompetence of the hospital has to be part of our assessment.'

In another similar case, the possible admission of a patient who had been violent in his local hospital was being discussed. All the Broadmoor consultants thought that he needed medium secure conditions. However, the medium secure unit staff thought he should stay where he was. The patient had been confined to a side room for almost a year. The view of some Broadmoor doctors was: 'It's not for us to ameliorate his conditions by admitting him, when he should be elsewhere. We should publicize the failure of the medium secure unit to take him.' Others disagreed, believing the primary consideration to be the patient's predicament: to take other matters into account 'would be using a patient as a political pawn to his own detriment.' The division of opinion illustrates how large a role is played by chance in the admission process: whether a potential patient is or is not recommended for admission to Broadmoor depends in large measure on the person who happens to be on duty when requests for beds are made. Of course chance also enters into the picture at an earlier stage, since whether an application for a special hospital bed is made at all will depend upon the personal views of the National Health Service consultant (or prison medical officer) in charge of the case.

In the cases mentioned so far, the potential patient was already assured of an NHS hospital bed, so the consultants knew that if they opposed acceptance, hospital care of another kind would be available. Often, however, the patient was in prison and Broadmoor appeared to its consultants to be the only way psychiatric care could be delivered. They assumed, even when the matter had not been put to the test, that seriously ill men who had committed crimes of violence would not be acceptable to National Health Service hospitals. Occasionally, mentally ill men who had been convicted of minor offences were recommended for admission to Broadmoor on the grounds that other hospitals had indeed refused to take them. In such cases it was usually suggested that Broadmoor should 'tidy the patient up' so that the local hospital might then be persuaded to take a controlled and improved patient. It will be seen from the sample we studied, that patients admitted in this way were nevertheless liable to remain in Broadmoor for very many years.

THE SAMPLE: LEGAL ADMISSION PROCEDURES

As explained in Chapter 1, our sample of men admitted under the legal classification of mental illness was not drawn from the total population of such men in Broadmoor. It was selected to match, on length of stay and legal admission category, Broadmoor's population of PD men. Since all the latter had been admitted as convicted offenders, this meant that mentally ill men who came to Broadmoor under civil sections or under the Criminal Procedure (Insanity) Act (1964) were not included in our sample.

The total number of men in our mental illness admission sample was 116. It represented about a third of the hospital's mentally ill male residents, and about half of its restricted male convicted offenders. The great majority (100) had been sent to Broadmoor on hospital orders with restrictions: in all but three cases the restrictions were without a time limit. Of the remaining 16 men, three were on hospital orders to which restrictions were not attached, 12 had been transferred to Broadmoor while serving prison sentences, and one man had been admitted from prison while on remand.

The research was carried out before the 1983 Mental Health Act came into operation, and all the men in our sample had been admitted under the 1959 Act. The hospital order provisions of that Act (sections 60 and 65) were broadly the same as those now contained in the Mental Health Act, 1983 (sections 37 and 41). Similarly, the provisions of the 1959 Act about the transfer of sentenced prisoners to hospital were very much like those now contained in the 1983 Act (section 47 and 49). They empowered the Home Secretary to order a prisoner to be transferred to hospital, if he was satisfied by two medical reports that transfer was necessary on the grounds of mental illness, psychopathic disorder, or mental handicap (called mental impairment in the 1983 Act). The Home Secretary also had the power to add a restriction direction to the transfer order: the effect of this was to prevent the prisoner from being moved from hospital without the Home Secretary's consent. Such restrictions cease to have effect when the prison sentence expires (in the 1959 Act, this was defined as the prisoner's latest day of release but the 1983 Act uses the earliest date of release for this purpose). Once the restrictions on the release of a transferred prisoner cease to have effect, he is regarded (under both the 1959 and 1983 Acts) as being detained under a hospital order without restriction. Although his sentence has expired, the man will thus remain under detention until either his consultant or a Mental Health Review Tribunal decides otherwise.

Of the twelve men in our sample who had been admitted as sentenced prisoners, four were life prisoners whose restrictions were therefore indefinite and eight were serving determinate sentences but had passed their latest date of release when we interviewed them. They were therefore being detained on notional hospital orders, and were no longer subject to restrictions. Thus, of the 116 men in our sample, the great majority, 104 (90 per cent), were, at the time of our research, restricted patients who could not be moved without Home Office permission. Twelve men (10 per cent) were not subject to such restrictions and were thus detained solely at the discretion of the hospital.

THE ADMISSION OFFENCE

What offences had the men committed? Details of the main conviction preceding admission are provided in Table 2.1; it includes the twelve men transferred from prison after sentence. It will be seen that the largest single category of offence was homicide, accounting for 33 per cent of the group. Other offences of serious violence had been committed by a further 35 per cent of men, and if these offences are combined with rape and arson to represent serious or potentially life-threatening behaviour they account in total for 80 per cent of the group. About three-quarters of the 116 men would have been legally eligible for a life sentence for the offences at admission.

Given the level of security provided by Broadmoor, it might be expected that all admissions under the criminal provisions of the Mental Health Act would be preceded by a life-threatening offence. However, this was not so for about a fifth of our sample. A variety of reasons were relevant. In some cases, the occasion of a man's appearance in court, albeit not on serious charges, represented for other agencies a welcome opportunity of relieving themselves of a burden. The following example is an illustration. For many years before the commission of his Broadmoor offence, the behaviour of this schizophrenic patient had given his local hospital cause for concern. He was threatening to staff and other patients and attempts were made to have him transferred to Broadmoor. The Department raised questions about the need for this, and while discussions were in progress, the patient was found in possession of a dangerous weapon and was reported to have stated that he intended using it. However, the only charge which could be brought against him was one of larceny. In ordinary circumstances it is unlikely that the offence (itself a technicality) would even have been proceeded with, but it offered the local hospital authorities the opportunity of having the patient removed. He was therefore charged, and when convicted of theft, was sent to Broadmoor on a hospital order with restrictions.

The commission of a relatively minor offence can thus facilitate a patient's removal to Broadmoor. A similar process may occur when the mentally ill offender is a former patient of the National Health Service living in the community. Recurrence of his illness would normally lead to his readmission to the hospital. But if he is detained by the police after an offence and then remanded in custody, the local hospital, finding that an unwanted client has been diverted from its care, may be reluctant to resume responsibility. Broadmoor may then become the only hospital to which the admission of an obviously ill man can be arranged.

In these ways, there is a group of mentally ill patients who are rejected by the NHS hospitals that know them and are admitted to Broadmoor via the criminal courts, even though their offences are not serious. They are really disguised civil admissions.

In a number of these cases, the DHSS had initially refused admission. In one, the patient, a chronic schizophrenic man, had been in a local hospital for ten years and had caused frequent difficulties because of his aggressive behaviour. After an attack on a nurse, the police were called; he was charged with assault and remanded in custody. A special hospital consultant came to examine him. The patient, he reported, needed 'the kind of long-term supervision that used to be available in the locked wards of conventional hospitals'. Since such care was not available in the National Health Service, and since the man was patently ill, the consultant believed he should be admitted to Broadmoor, even though maximum security was not required. The consultant's report added: 'I would expect he could be transferred back to a conventional hospital after a year or two.' The DHSS eventually agreed to accept him. Seven years later, although his controlled behaviour in Broadmoor had been repeatedly noted, no attempt had been made to move the patient. The case illustrates what is, as we shall see later, a crucial facet of the admission process: even if the admission is a 'marginal' one, once the patient is inside Broadmoor many factors, including inertia, can combine to keep him there.

Almost a fifth of our sample, 22 men, were in-patients of the NHS when they committed the offence that led them to Broadmoor. Their offences were less serious than those of other mentally ill men. When we rated the violence level of offences, using a scale developed by Gunn and Robertson (1976), we found that the maximum rating (offences causing death or serious injury) was given to one-third of the National Health Service men, as compared with two-thirds of the others ($x^2 = 6.51$ df $p < 0.02$). The proportion of men with admission offences of dishonesty was also significantly higher in the National Health Service group.

Where offences of violence had been committed, the victims were, in 35 per cent of cases, members of the man's family, usually close relatives, 14 per cent being parents and 13 per cent wives. Strangers accounted for 27 per cent of victims. The only other numerically significant categories were 'friends' and 'hospital staff', each of which accounted for 12 per cent of cases. The victims of violent offences were evenly distributed between the sexes. The age distribution of victims was wide, ranging from 1 year to 83 years; 14 per cent of victims were under 18 and the same percentage of victims were over 60 years of age.

Irrespective of the nature of the conviction, we noted whether any sexual assault had taken place or been attempted at the time of the offence. Only seven men (6 per cent) had committed such an assault, and a further seven men appeared to have intended a sexual assault which was frustrated.

PREVIOUS CONVICTIONS

Details of previous convictions are presented in Table 2.1.* The majority of the men (61 per cent) had a previous criminal record, but only a minority (30 per cent) had ever served a custodial sentence. It was of interest to note that 82 per cent of the men admitted after a homicide offence had had no previous conviction for violence.

As regards their criminal histories, the men in our sample were very different from men in the prison population. Not only were there many more first-time offenders and men who had never been in prison, but the pattern of offending was different (Gunn *et al.*, 1978). Only 19 men (16 per cent) had been convicted as juveniles; 44

* All tables appear in Appendix II at the end of the book (pp. 141–63).

per cent had no convictions before they reached the age of 25, the mean age of first conviction for the group. In these respects the men not only differed sharply from prison populations but were, as we shall see, almost identical with samples of mentally ill offenders in local hospitals. It seems that the offending behaviour of many was related to the onset of their illness.

PSYCHIATRIC DATA

Data about the men's previous contact with psychiatric services are presented in Table 2.2. It will be seen that two-thirds of the men had been admitted to a psychiatric hospital on at least one occasion before coming to Broadmoor. This proportion is very similar to that reported by Walker and McCabe in their study of all hospital-order patients in a twelve-month period in 1963/64 (Walker & McCabe 1973). The average number of previous psychiatric admissions for the population was a little over 3, but the variable was very widely distributed.

For a hospital order to be made by a court, or for a prison transfer to be effected, it is necessary for two suitably qualified doctors to state that the person concerned is suffering from either mental illness, psychopathic disorder, subnormality, or severe subnormality (now referred to as mental impairment and severe mental impairment under the 1983 Mental Health Act). The term mental illness was undefined in both the 1959 and 1983 Acts. The court form requires the doctor to present evidence on two points:
1. Information to establish mental illness, including reference to kind of illness and description of symptoms.
2. Reasons for the conclusion that the mental illness is of a nature or degree which warrants the detention of the patient in hospital for medical treatment.

We examined the medical records of the men in our sample to determine the diagnosis given them at the time of their trial or when their removal from prison was requested. The great majority were categorized as schizophrenic: diagnoses of schizophrenia and paranoid schizophrenia had been given to 52 per cent and 37 per cent of the group respectively. The way in which the various reporting doctors used the two terms in their reports is not a matter on which we had information, but it seems likely that there was considerable overlap. Nevertheless, when we came to analyse our data by these diagnostic categories, we found, as will be seen, some significant differences between them.

Only 13 men were not described at the time of their trial as either schizophrenic or paranoid: six of these were diagnosed as suffering from affective psychotic disorders, three from other psychoses, one from an organic condition, and three from non-psychotic depression. Two of the latter were by the time of the research regarded by the Broadmoor doctors as personality-disordered rather than mentally ill. One was a young man whose extremely violent offence had been found difficult to explain by all concerned. He had shown no previous evidence of violent behaviour and had led an otherwise con-centric existence. The case which was made out for his having suffered from a depressive illness was extremely weak from a clinical standpoint: after examination in prison and in the context of an offence which involved multiple killings, the best evidence for depression presented in the reports stated that the man had 'clearly recovered a good deal . . . but is still somewhat preoccupied and unable to show any interest in wider topics such as his future career.' The psychiatric reports in this case provided excellent examples of what Glanville Williams (1978) has referred to as 'a benevolent con-spiracy between psychiatrists and the Court' to ensure that action is taken, in this case admission to a special hospital, which both consider desirable.

DIFFERENCES BETWEEN DIAGNOSTIC GROUPS

The psychiatric, social, and criminal data we had collected were analysed by the diagnoses given on the trial reports. For this purpose, the population was divided into three subgroups: the first comprised 43 paranoid schizophrenic men, the second consisted of 60 other schizophrenics, and the third was of 13 men with diagnoses other than schizophrenia. The statistically significant differences produced by these comparisons are detailed in Tables 2.2 and 2.3.

Group differences concerning psychiatric variables are presented in Table 2.2. It is clear that many more men in the undifferentiated schizophrenic group than in the paranoid subgroup had had a previous psychiatric admission, and that the average age at first admission was considerably younger for the undefined schizophrenic group. Paranoid schizophrenia is associated with and defined by relatively intact functioning, and it may be this association which accounts for the group difference in age at first admission to hospital, rather than later onset of the illness. Overall, the pattern was for the undefined schizophrenic group to demonstrate the greatest degree of social and psychiatric disability, and for the non-schizophrenic group to be least stigmatized by illness and illness-related characteristics: more of them, for example, were working at the time of the offence, and more of them had been married.

Table 2.3 shows that, with one exception, the subgroups did not differ from one another on pre-admission criminal data. The exception was age at first conviction, the mean age for the undefined schizophrenic group being the lowest (25.5), and for the non-schizophrenic group, the highest (33.2). It can be seen from Tables 2.2 and 2.3 that, in each diagnostic group, mean age at first conviction preceded mean age at first admission to hospital by about one year. These data suggest that whatever the diagnosis, there may have been a causal link between the emergence of the illness and the man's first conviction.

The groups also differed in respect of the admission offence, some details of which may be found in Table 2.3. Two points were of interest. First, the paranoid schizophrenic group tended to contain a higher proportion of homicide offenders than did the other schizophrenic group (37 per cent as compared with 23 per cent; $\chi^2 = 2.33$ 1 df $p < 0.15$). Secondly, there was a suggestion of an inverse relationship between the severity of the illness and of the offence. As we have seen, in terms of disability, the undifferentiated schizophrenic group was the most severely ill, and the non-schizophrenic group the least severely so. In terms of their admission offences, the former group had the highest proportion of men who had not hurt anyone, whereas all the men in the latter group had committed severely violent crimes. The main interest of the relationship is that it shows that there are a number of different routes into Broadmoor, and that even within this highly selected offender population there is considerable variation in respect of both illness and degree of violence exhibited.

MAJOR CHANGES OF DIAGNOSIS IN BROADMOOR

When we interviewed them, we asked the consultants to give us their diagnosis for each man. Of the 116 men admitted under the legal category of mental illness, the doctors in nine cases (8 per cent) gave us a diagnosis of psychopathic disorder and said that they did not consider the patient to be suffering from mental illness. Only one of these nine men had been formally reclassified under the Mental Health Act.

In Appendix I, we give a detailed account of these nine men. In this, it will be seen that each case was different, but three elements tended to recur. The first was the presence of obsessional rigid thinking combined with abnormal sexual drive. Obsessional behaviour can be the precursor of a schizophrenic illness and for this reason it was not surprising to find that some of these men were categorized as schizophrenic. It may well be that, without the need to produce a hard diagnosis for the court, the formulation of schizophrenia would not have been made, or would have retained a significant question mark over it.

The second recurring factor was that of deliberate deception: four of the nine men said they had originally either faked or exaggerated symptoms of mental illness, usually in order to avoid imprisonment. Their years in Broadmoor under constant observation had certainly not revealed any signs of illness. Finally, the third element which was obviously common to all the cases was the conviction of the doctors that, whatever the niceties of diagnosis might be, admission to Broadmoor was the most appropriate and desirable disposal for the case. This might be for various reasons: perhaps because no other outcome could guarantee long enough detention, or because alternative disposals seemed inappropriately punitive. It is worth noting here that, although the evidence for a diagnosis of mental illness was very weak in some instances and in others depended on the offender's success in deceiving the doctors, all these cases could, had the doctors so wished, have comfortably been accommodated in the legal category of psychopathic disorder. The following case illustrates the point. The offender was a youth who had attacked a stranger: he had no history of violence. In the remand prison he told the doctors that he had been driven to the attack by voices telling him to kill. No psychotic phenomena were observed during the remand period, but the medical reports, including one from a special hospital doctor, described him as suffering from schizophrenia: they recommended admission to a special hospital on the grounds that his illness made him dangerous. After six months in Broadmoor the consultant who had seen him on remand reported that further observation was necessary before a final diagnosis could be given. 'He has clearly been an unreliable witness . . . He now contends that the voices commanding him to kill . . . were pure fabrication (designed) to secure committal to hospital rather than imprisonment . . . He displays highly irresponsible and dangerous conduct to such a degree as to warrant his placement in the psychopathic disorder category if the diagnosis of schizophrenia cannot be sustained'.

THE SCHIZOPHRENIC SPECTRUM: THREE CASES

As we have seen, diagnoses of schizophrenia and paranoid schizophrenia accounted for 89 per cent of our sample. In order to give the reader a fuller understanding of the range of illness managed in Broadmoor, we describe below three patients whose histories reflect the breadth of phenomenology subsumed under the label of schizophrenia.

The first was chosen because his history and presentation indicated that his illness had left him intellectually intact and, in regard to most of his functioning, quite normal. His illness became apparent only when his systematized delusional beliefs were explored. He had been in Broadmoor for 10 years when we saw him, and had had one previous admission to a psychiatric hospital before committing his offence. He was extremely articulate and of above average intelligence. When he was interviewed, his conversation was appropriate and unremarkable until the nature of his offence was discussed, when it became clear that he believed his homicidal behaviour at that time to have been reasonable and fully justified. The delusional beliefs which had then guided his actions were with him still and, it seemed, in equal measure despite a decade of phenothiazine medication in Broadmoor. In this case, the hospital was providing security for a man who was obviously dangerous and whose illness had shown no response to treatment.

A far greater number of the mentally ill men in our sample were suffering from an illness which presented itself in a more general way than that described above, and which affected feelings, thinking, and behaviour over a wide range of activities. The man chosen as a representative of this group was in his mid fifties when we saw him and had been in Broadmoor for more than 15 years. Before his admission he had been subject to prolonged bouts of mental illness, these having started when he was in his late twenties. He had been admitted to hospital on more than a dozen occasions and had been treated with phenothiazine medication to which he responded well. However, once well, he was in the habit of absconding from hospital and, finding himself homeless, would quickly get into trouble. His admission to Broadmoor under a restriction order was precipitated by a conviction for larceny. The police had found him wandering abroad, floridly psychotic, with an implement which could have been used as a weapon. He was prosecuted and sent to Broadmoor, a hospital from which he could not abscond. For a number of years before his admission he had been unable to sustain himself outside hospital without recourse to theft and burglary. He was admitted to Broadmoor because the local hospital felt it could no longer contain him and because the combination of his breaking and entering activities with his mental illness, when unmedicated, made him potentially dangerous. When we saw him for our research, his illness was in remission in the sense that no florid or obvious disorder of thought processes was evident. The patient's stay in Broadmoor had been uneventful and he told us he had no wish to be transferred. For 15 years Broadmoor had provided maximum security for someone subject to an illness for which there was relief but no cure, and whose behaviour and life-style, in combination with that illness when untreated, were thought to make him dangerous.

It was the absence of symptom-free periods that characterized the condition of our third patient. He was selected to represent a group of men whose illness was marked by a gradual insidious deterioration of their mental functioning that resulted in the disintegration of their personal and social behaviour. The effect of their illness was to render them debilitated and in need of care. The middle-aged man chosen to represent this group had been admitted to a local psychiatric hospital in his late teens and had subsequently been readmitted there on three occasions. The admissions were all for long periods and he had never entered into employment. The behaviour which brought him to Broadmoor was his tendency to light fires in inappropriate places. After one such incident, his local hospital decided to take legal action and his transfer to Broadmoor was effected via the court. His illness responded to phenothiazine medication only in so far as his behaviour was repressed by such powerful tranquillizing agents. The inadequacy of his mental functioning remained unchanged by drugs and by ECT; neither made him well in the sense of restoring his thinking to some sort of order. Although he was admitted after a criminal offence, he was in reality a debilitated man who came into and remained in Broadmoor for want of a better alternative and not because he required the maximum security offered by its walls.

PLACING THE MENTAL ILLNESS GROUP IN CONTEXT

Thus far we have described the characteristics of the mentally ill men in our study without reference to any other group. In the present section we attempt to place our population in a broader psychiatric context. We were fortunate in having access to data collected from a wider survey of offender patients and these are presented in Table 2.4. The figures have been culled from a follow-up study by Robertson and Gibbens (1979) of the hospital-order population collected originally by Walker & McCabe (1973). The right-hand part of the Table presents details of men categorized as mentally ill and sent to local psychiatric hospitals under restriction orders in the years 1962, 1963, 1964, and 1975.

It can be seen in Table 2.4, that the Broadmoor and local hospital groups are similar in respect of previous psychiatric admissions and psychiatrically related criminal variables such as age at first conviction. In other respects too, the groups were alike. Almost identical percentages had a criminal record before the admission offence (59 per cent of the Broadmoor men as compared with 62 per cent in the local hospital population), and in terms of demographic variables few differences emerged.

There were however two major differences between the groups. One was in the distribution of diagnostic categories. Over a quarter of the men admitted to local hospitals were described as suffering from affective illnesses as compared with only 8 per cent of such men in the Broadmoor group. On the other hand, among the Broadmoor intake there were significantly more men in the paranoid schizophrenic category. The second difference, overwhelming all others, is the enormous discrepancy in the nature of the admission offence. Almost half (47 per cent) of the Broadmoor sample had been convicted of homicide, but this offence accounted for only 7 per cent of the local hospital group. It is thus the nature of the offence which determines whether restricted offenders are detained in local hospitals or sent to Broadmoor. The exception to this general rule is the depressive homicide offender, who is likely to be sent to a local hospital. However, in both local and special hospital, the majority of restricted patients have diagnoses of schizophrenia, paranoid or otherwise. The destination of offenders with these diagnoses is determined by whether their attack on another person resulted in the other's death.

3 Treatment of psychotic men

INTRODUCTION

Information was collected about the treatment received by the men, and what they and their doctors had to say about it. The data were gathered from the case notes, prescription sheets, medical records, and from our interviews with patients and doctors. The records showed what physical treatments had been used (i.e. medication or ECT) but were often deficient in information about psychotherapy or other non-physical treatments, and on whether the patient's suitability for these had ever been assessed. On such matters it was frequently the patient himself who had the fullest information.

For the purpose of discussing treatment, we are including in our group of mentally ill men, all patients diagnosed as psychotic by their doctors at the time of our research, irrespective of their legal classification on admission to Broadmoor. Twenty men admitted with a Mental Health Act classification of psychopathic disorder were diagnosed as psychotic by their consultants when we interviewed them: these men are included among the mentally ill in our analyses, even though only three of them were formally reclassified as such under the Mental Health Act. Conversely, the nine men who came to Broadmoor with a Mental Health Act classification of mental illness but were not diagnosed by the Broadmoor doctors as mentally ill (see Appendix I) have been excluded from the present chapter. Thus 127 men diagnosed by their consultants as suffering from psychotic illnesses comprise the group discussed here; we refer to them in the text either as mentally ill or as psychotic. Only five of these men declined to be seen; another man was too ill, and four were moved from Broadmoor before we could interview them.

DIAGNOSTIC GROUPS

Details of diagnoses are shown in Table 3.1. All fall within the range of psychotic disorders listed in the Ninth Revision of the International Classification of Diseases (World Health Organization 1978). We decided, for the purpose of examining the data on treatment, to divide the men into three groups: the paranoid schizophrenics (46 men), the other schizophrenics (65 men) and the remaining group of 16 men, most of whom were suffering either from some organic condition or from a bipolar affective disorder. Table 3.2 shows some of the main differences between the groups but it should be noted that there was no difference in the time they had spent in Broadmoor, 8.5 years being the average length of stay for each.

TREATMENT BY MEDICATION

In any psychiatric hospital, nursing care and medication are the main treatment offered to the mentally ill, and Broadmoor is no exception: 89 per cent of the men were on psychotropic medication when we interviewed them. The proportions differed between the three diagnostic groups: 96 per cent of the paranoid men, 89 per cent of the other schizophrenics, and 69 per cent of the third group were on such drugs when interviewed. Table 3.2 shows the details and the drugs being used, and Table 3.3 shows how much in the way of major tranquillizers was being prescribed for each patient at the time of the research. In the Table each patient's dose has been converted into units equivalent to 300mg of chlorpromazine daily. Conversion data were provided by Aschkenasy and Carr (1982), who had calculated equivalent values on the basis of an extensive literature search and clinical experience of patient response to titrated measures of the drugs involved. Almost half (45 per cent) of those on major tranquillizers were receiving more than one variety. (For a comparison with local hospital practice see Edwards & Kumar 1984.) Eighty-three per cent of the men on major tranquillizers were also on anti-Parkinsonian drugs.

At the time of our research, the provisions of the 1983 Mental Health Act about consent to treatment had not been introduced, nor was there any legal requirement for a detained patient to be informed about the nature, purpose, and likely effects of the treatments he was to be given. When we asked the medicated patients what their drugs were for, almost half of the paranoid men (47 per cent), as compared with 23 per cent of other schizophrenics and 11 per cent of the non-schizophrenic group, said that they did not know ($\chi^2 = 8.41$ 2df $p < 0.02$). The paranoid men also expressed least satisfaction with their medication. Only a third thought it had helped them, whereas two-thirds of the other schizophrenics and all of the non-schizophrenic group thought their drugs had helped. Not surprisingly, it was the paranoid men who most often wanted to come off their medication.

The consultants were asked about the role of drugs in treatment: they thought them important or very important for over 90 per cent of the men in each of the three diagnostic groups. But the limited efficacy of medication, especially for the schizophrenic groups, was evident: despite treatment, over half of these patients were categorized by their doctors as deluded (Table 3.2).

TREATMENT BY ECT

None of the mentally ill men was undergoing a course of ECT treatment when interviewed, but almost half (49 per cent) had received such treatment in Broadmoor: the proportions did not vary significantly between the diagnostic groups. It was often unclear from the records exactly why ECT treatment had been administered, but it seemed that in about a third of the cases it was because of the patient's depression, and in half the cases it was an attempt to control psychosis. Table 3.4 shows the number of ECT courses which the treated men in the three diagnostic groups had received during their stay in Broadmoor, and the total number of treatments. The non-paranoid schizophrenic group had received the greatest number of treatments.

The men who had received ECT were asked whether it had helped them. Fewer than half (40 per cent) thought it had, and the proportions did not differ significantly between diagnostic groups. The Broadmoor records contained little systematic information about the results, beneficial or otherwise, of ECT treatment, and could not be used for the purpose of measuring its efficacy.

NON–PHYSICAL TREATMENTS

Non-physical treatments such as psychotherapy or social skills training were not as a rule given to psychotically ill patients. Only 39 men (31 per cent of the sample) had ever in the course of their stay participated in treatments of this kind, and at the time of the research no more than ten men (8 per cent) were engaged in them. We asked the men who had ever been involved in these treatments to say how helpful they had found them. Seven men had received individual psychotherapy at some point, on average for a period of 9 months (s.d. 7.9). Three had found it very helpful, three had not found it useful at all. Twenty-seven men (21 per cent) had attended group therapy at some point, on average for a period of 20 months (s.d. 21.5). Over half of these men (56 per cent) said they had not really found the experience helpful; a third had found it quite helpful, and two men (8 per cent) said the treatment had been very helpful to them.

Three patients had received behavioural modification treatment or relaxation therapy from psychologists: only one of these men had found these sessions helpful. Another eight men had received other forms of help with special problems, usually from the speech therapist: most had found it helpful. Finally, there were seven mentally ill men who had taken part in social skills courses and another four who had been to a sex education course. Half of these patients had found the courses helpful and half had not. The former included a poorly socialized youth who had found the social skills course a revelation: 'It's taught me things I never dreamt of'.

INTENSIVE CARE AND INCIDENT REPORTS

When we examined the case notes and hospital records, a note was made of any incident forms recording violent or undisciplined patient behaviour. Obviously not every single incident will have been recorded in this way, but the records provide a fairly accurate picture of the amount of serious disturbance caused by each patient. Thirteen per cent of the men had been recorded as having been involved in self-injury of some sort, 28 per cent as having been involved in some kind of violence against the staff, and 43 per cent as having been involved in a disturbance with another patient. There was no difference between the diagnostic groups as regards the number and type of incidents recorded. Among the mentally ill group who had, on average, been in Broadmoor for eight and a half years, the mean number of recorded violent incidents against staff was 0.83 (s.d. 2.27) and of violent incidents with fellow patients was 1.9 (s.d. 3.8). However, as reflected by the large standard deviations, a few men were in trouble again and again. An example was the young and very disturbed schizophrenic man who was involved in ten fights with other patients, and six incidents with nurses during the first five years of his stay. In addition he repeatedly attempted suicide and self-injury. Despite treatment with medication and ECT he was, in his own words, 'always in and out of the punishment block'—the patients' name for the intensive care unit.

About 40 men were housed in the intensive care unit at the time of the research. In the main, it was the disturbed, impulsively violent psychotic men who were sent there, usually after a violent incident. A third (35 per cent) of the mentally ill sample had been in this unit at some time during their stay. Three-quarters of those who had been involved in violent incidents against staff had been there, as compared with 39 per cent of the men who had been aggressive against other patients but not against staff.

A number of men commented on their stay in this unit. Some found its highly structured routine congenial and did everything they could to stay there. Others commented negatively on the lack of activity, 'sitting in a chair staring at the wall all day', and on the controlling nature of the regime.

CONSULTANTS' VIEWS OF THE RELATIONSHIP BETWEEN TREATMENT AND OFFENDING

All the men in our sample had been sent to Broadmoor after a criminal conviction. We asked the consultants in respect of each of their patients: what causes him to offend, and what in your opinion can be done about it? The answers were usually very clear-cut. The doctors said it was the illnesses of these men that had caused them to offend. As a rule this was because delusions or hallucinations had led them into committing acts of violence, but occasionally it was because the illness was thought to have disinhibited them. In either case, the consultants thought that the remedy for the offending was clear: treatment of the illness with medication.

CONSULTANTS' VIEWS ABOUT TREATMENT

The doctors were in no doubt about the need of the psychotic group for treatment. Virtually every patient was said to have needed hospitalization when first admitted, and 92 per cent were rated as definitely needing psychiatric treatment at the time of our research.

The consultants were asked to categorize the mental state of each patient in terms of improvement since admission: they rated 60 per cent of the mentally ill men as improved, 18 per cent as much improved, 20 per cent as unchanged, and 2 per cent as deteriorated. There was a trend for the non-paranoid schizophrenics to be rated as 'unchanged' more often than other patients (30 per cent were so rated, as compared with 15 per cent of the other two groups $(p < 0.19)$. For patients whose condition was said to have improved, we tried to establish from when the improvement dated. We were rarely able to obtain information on this from the doctors, or to find it in the case notes.

We asked the consultants about the contribution that treatment, as opposed to the passage of time, could be expected to make in rendering each man well enough to be moved from Broadmoor. For 63 per cent of the men, the doctors thought that treatment would be the major element. For another 16 per cent, they thought that the roles of time and treatment would be equally important. In 10 per cent of patients, the doctors thought that the effects of time rather than of treatment would in the end enable the patient to leave Broadmoor: the hope was that the illness, though not responding to medication, would eventually burn out. For another 11 per cent of the psychotic men, the consultants feared that neither treatment nor the passage of time would bring about enough improvement to enable the patient to be released from Broadmoor. Some men in this group had brain damage, some had multiple diagnoses that made the prognosis poor, and in some it was the intractability of the schizophrenic illness that blighted the patient's prospects. An example from this last group was a man who had been in hospital since his adolescence and who came to Broadmoor in his early thirties, when his local hospital could no longer manage his uncontrolled behaviour. By the time of the research he had been in Broadmoor for 7 years, and remained floridly psychotic despite treatment with drugs and ECT. He was repeatedly involved in impulsive violence against staff and patients, and was frequently sent to the intensive care unit. His consultant could see little hope of improvement, pointing out that the severe illness had remained unabated despite twenty years of continuous treatment.

We asked the consultants how important they considered the role of long-term care in Broadmoor to be in the treatment of each case. For 42 per cent of the men they thought it had no important role to play. But for the majority of men they rated long-term Broadmoor care as either important (21 per cent) or very important (36 per cent).

Patients for whom long-term care or asylum in Broadmoor was thought by the doctors to be important or very important were compared with other psychotic men on the data we had collected. The main difference that emerged related to the severity of illness. The long-term group were those who had not responded to medication, who were more often rated by the consultants as deluded and hallucinated, who had been involved in more violence against the staff, and who had more often come to Broadmoor from other hospitals. An apparently paradoxical difference between the groups was that the offences of the long-term group were less serious than those of other patients. This was because men who had proved unmanageable in local hospitals were over-represented in the long-term group; as noted earlier (Chapter 2) these men often came to Broadmoor because of the severity of their illnesses rather than of their offences.

BROADMOOR AS AN ASYLUM

Not only the doctors, but some of the psychotic men themselves saw Broadmoor as an asylum. They had experienced what their consultants had observed: that local hospitals find ways of rejecting difficult patients, however ill they may be.

One-fifth of the mentally ill men had come to Broadmoor because other hospitals had sought their transfer, and others had come after local hospitals had declined to accept them. Such men were well aware that they were unwanted rejects from ordinary psychiatric hospitals, and the appreciation that some of them expressed for Broadmoor's asylum was poignant in the circumstances: 'I couldn't have got by without Broadmoor . . . I couldn't have got by without the drugs . . . I'd have ended up in the gutter . . . The trouble with (my local hospital) was that they would chase me out'. These were the words of a paranoid schizophrenic man who had been in Broadmoor for eleven years, and remained very ill.

The concept of Broadmoor as an asylum or haven was voiced by a number of other patients. In the view of one schizophrenic man, it offered 'a little bit of peace and no real pressure'. Another appreciated it as 'a breathing space with unlimited time to benefit from treatment'. A manic-depressive patient who had suffered much rejection in the National Health Service and who now felt stabilized on medication said: 'You get help when you really need it . . . it's given me time to come to terms with my illness, to learn how it works and how to cope with it'.

Some measure of the extent to which the patients felt that they were receiving welcome asylum was obtained from their answers to our question: 'Do you feel ready to leave Broadmoor?' Almost a fifth (18 per cent) of the mentally ill said categorically that they did not feel ready. The proportion was highest in our third diagnostic group, where five out of 13 men who answered the question (38 per cent) said they did not feel ready. There was no relationship between readiness to leave and length of stay. The most usual reason given for not wanting to leave was the fear of not being able to cope elsewhere or the feeling that they had become settled and content in Broadmoor. 'I don't think I am missing that much outside . . . it's not so great there', is how one young man put it.

Obviously this is not a reason for staying in Broadmoor that is acceptable to official DHSS policy. The Department does not condone keeping patients there for any reasons other than their need for maximum security. Among the men who told us they did not want to leave the hospital was a patient in his seventies who had been in Broadmoor for a decade. His local hospital was prepared to accept him, but was reluctant to do so, pointing out that conditions there would be worse for the patient in terms of staff attention and support. The Broadmoor consultant agreed that this was so, and the patient himself was adamantly opposed to any move. The DHSS was reluctant to accept that so elderly a man should be held in a special hospital only on humanitarian grounds, but in the event the patient was not moved. The case points to a lack in the National Health Service of good long-term care or asylum facilities for those who are chronically and seriously mentally ill: local and Broadmoor consultants were agreed that conditions for this patient would have been inferior in the National Health Service. It is ironic that, because Broadmoor is one of the few places where good asylum is available, it should be given there to any patient who has no need to be in maximum security.

PATIENTS' OPINIONS ABOUT TREATMENT

The men's opinions of the specific treatment they had received, medication, ECT, psychotherapy, were reported earlier in this chapter. But the experience of being in Broadmoor obviously consists of many other things, such as living together with other Broadmoor patients, and being looked after and locked up by Broadmoor staff. What did patients have to say about it?

(a) Satisfaction with treatment

We asked a number of questions to try to elicit their views. First, were they satisfied with their treatment? Over half (54 per cent) said they were, and the proportions did not differ significantly between the diagnostic groups. About 12 per cent were ambivalent, satisfied in some ways and dissatisfied in others. A substantial minority (27 per cent) expressed unambiguous dissatisfaction.

The most common cause of dissatisfaction among the patients was the feeling that they had been overlooked, and that nobody was giving active attention to their case and to their need still to be in Broadmoor. The feeling was expressed in various ways. Men said they had not seen enough of their doctors; that they felt forgotten; that there was no definite programme directed towards helping them to move out; that their problems had not been adequately assessed or treated. Many comments were made about the paucity of contact with consultants, in reporting which we recognize of course that the work consultants do for their patients is by no means all done face to face. One patient who was full of praise about many aspects of the treatment, deplored how seldom the men in his ward were seen by the consultant. He added 'But I've been lucky, I've been in trouble, so I see him more often!' It should be noted that it was not possible to use entries in the hospital case notes for the purpose of examining how often patients were seen by their consultants, for more often than not these were inadequately sparse. In some cases, years went by without any psychiatric information being recorded in the notes.

The population, as we have seen, was a long-term one, and because of staff changes and moves, the consultants had often not had continuing care of their patients: half of the mentally ill men had been under their present consultant for less than three years. We asked the consultants how well they knew each of their patients, and we asked the men how well they thought their consultants knew them. The replies showed a considerable degree of discrepancy. The doctors thought they knew their patients well in 73 per cent of the cases, but only half of the patients thought their doctors knew them well. Forty per cent of the patients thought that their consultants did not know them well, but in only 27 per cent of their cases did the doctors say that they did not know the patient well.

It was thus clear that in a substantial proportion of cases, the men were unhappy with this aspect of the patient/doctor relationship. A smaller, but still substantial proportion of the consultants also expressed dissatisfaction about it. It should be said that the importance of the consultant knowing his patients is greater in Broadmoor than it is in an ordinary psychiatric hospital, for it is effectively only through the consultant that the restricted patient can get out. The 1983 Act now empowers tribunals to discharge restricted patients, but it still remains true that the great majority of them will only be discharged if their consultants recommend this. When we discussed with the consultants how they arrived at their discharge decisions (Chapter 4), one point emerged clearly: it was not until the doctor felt he knew the patient and his case really thoroughly, well enough to anticipate and answer confidently any objections that might be forthcoming from a Minister's Private Office, that he was prepared to recommend release from Broadmoor. So the fact that a considerable number of men were not well known to their consultants constituted for those men a bar to movement out of the hospital. In one case, for example, the patient had been in Broadmoor for seven years. He was a schizophrenic man, who had committed a dangerous offence when untreated and acutely ill. With medication he became quiescent and presented no management problems. He had been under the care of the same consultant for most of his stay, and when we asked this doctor about suitability for transfer, he said he really did not know the man's present condition well enough to say: he would need to gather more information. When we asked the patient why he thought he was still there, the reply seemed accurate enough: 'I seem to have slipped by without them noticing'. In this way, to use the Broadmoor phrase, patients got lost in the woodwork, and it is not surprising that they expressed unhappiness about it.

(b) Had Broadmoor helped?

We asked the men whether they thought Broadmoor had helped them. Over half, 55 per cent, said that it had, a quarter said it had not, and eight per cent were uncertain. Those who said they had been helped were asked in what way. The most common reply, given by a quarter of the psychotic men, was that it had helped with their illness. Twelve per cent said Broadmoor had helped them to learn more about themselves, eleven per cent said it had taught them better self-control, and seven per cent said that it had helped them to get on better with other people.

An example of a patient who told us very positively that he had been helped was a young man who had been admitted five years previously, after a sexually motivated offence. There was some doubt about his diagnosis, but his consultant regarded him as schizophrenic and he was on anti-psychotic as well as anti-depressant medication. The patient found this beneficial: 'I'm less worried, it calms me and helps me to talk'. He was one of the very few mentally ill men who at the time of the research was taking part in a treatment programme that included individual and group therapy as well as social skills training. He found all these valuable, and also commented favourably on his contacts with some, but not all, of the nursing staff, whom he described as 'very helpful . . . they understand a lot'. He felt Broadmoor 'has given me a chance in life: it's taught me to cope . . . I'm overcoming my problems slowly, but surely'.

As in other hospitals, the concept of the therapeutic environment is part of Broadmoor's official ethos. It represents the view that therapeutic benefit is to be derived from all aspects of hospital life, not only from specific treatment procedures, but from the way in which patients are cared for by staff, and from activities like work, social recreation, and hobbies. It should however be said in this context that Broadmoor is not run as a therapeutic community in the technical sense.

We asked the consultants to rate how important they thought the therapeutic environment to be in the treatment of each man. They rated it as an important or very important element in the treatment of almost all (92 per cent) of their psychotic patients. However, when, at the beginning of our study, we asked patients about the different elements in their treatment, we found that they did not regard being in Broadmoor as a treatment in itself: treatment was seen as consisting of medication, or interviews with doctors. The replies on this point showed so little variety of opinion that we decided to drop the question from our interviews. Of course that is not to say that patients did not benefit from the non-specific part of their treatment in Broadmoor or that they were unaware of its importance. One patient told us that Broadmoor's chief value was that 'you learn to behave yourself and toe the line'. Another patient who told us how Broadmoor had benefited him was a young schizophrenic man who had been floridly psychotic at the time of his offence. It seemed that he had become ill some years before, but had never sought treatment and had lived in virtual isolation. In Broadmoor he was treated by ECT and medication, and made a good recovery. He told us how in the hospital community he was beginning for the first time to experience social life: 'When I left school, I never went out, never mixed, never worked. Now I'm seeing people, . . . mixing, . . . working. I'm more outgoing, . . . less shy. I didn't use to say anything. Now I'm coming out of my shell'. For this man, the experience of being in Broadmoor was genuinely therapeutic: both the specific treatment he was receiving (drugs) and the organization of hospital life were helping him recover.

(c)   Was it fair?

An offender consigned to a special hospital by the courts loses his right to a sentence based on considerations of criminal justice: doctors, not judges, will effectively determine the length of his detention in maximum security. We asked each patient whether he thought what had happened to him was fair. Some told us that 'fair' was not an appropriate term to use in their case as they did not feel they had been responsible for what they had done. However, most men did express an opinion about whether or not they had been fairly dealt with after their offence. Only a third of the mentally ill men said that they thought what had happened to them was fair. Half said categorically that it was not, these men being divided evenly between those who thought it unfair because they had been held too long, and those who made no such qualification. Although only a minority of the men felt that their detention in Broadmoor was fair, the majority (60 per cent) accepted that at the time of their offence they were ill and needed to be in a hospital. By being sent to Broadmoor however, they were prevented from going to their local hospital, and were channelled instead into one where they were held longer and under far more restrictive conditions than they would otherwise have experienced. When we asked whether they thought of Broadmoor as a hospital or as a prison only a third of the men (34 per cent) said that they regarded it as a hospital; the rest regarded it as a prison (30 per cent) or as a mixture of prison and hospital. It is thus not surprising that many felt that they had been unfairly treated.

In their views on what was fair, the men were also influenced by the ordinary considerations of criminal justice. Every one of the transferred prisoners who were being held after the expiry of their sentences, felt that they had been unfairly treated. Among hospital-order cases, patients whose offences could have attracted life sentences less often felt that they had been unfairly dealt with than did those whose offences had not been serious enough to be eligible for life imprisonment; 43 per cent of the former but 66 per cent of the latter felt that what had happened to them had been unfair ($\chi^2 = 4.32$ 1 df $p < 0.05$).

(d) Prison or Broadmoor

We asked the men whether, knowing what it was like, they would rather have come to Broadmoor or served a prison sentence. All had been in prison on remand. Thirty-six per cent opted for Broadmoor, usually giving its treatment and living conditions as the reason. Twelve per cent said they would choose Broadmoor for its conditions, but prison for determinacy of sentence: there was a widespread belief that in the prison system even lifers usually know how long they have to serve. The largest proportion of men (45 per cent), said unequivocally that they would rather have gone to prison than Broadmoor: again, belief in the determinacy of prison sentences was the reason most often mentioned.

(e) **Worst and best features of Broadmoor**

We asked what the men thought was the worst thing about being in Broadmoor. The most frequent reply centred on the loss of freedom, the restrictions, the constant control. In the local National Health Service hospitals with which most of them were familiar, patients are normally free to come and go, and they remain part of the ordinary world. Confinement in Broadmoor is a wholly different experience, and for most of the psychotic men, it was their first taste of prolonged incarceration. They talked of what it felt like to lose freedom, independence, and control; not to be able to choose what to do or what company to keep. In addition to the locked doors which kept them away from the ordinary world, the internal routines that went with maximum security were also heavily felt. One patient put it thus: 'when you go to work, or anywhere, you always have people to take you there . . . it's like a young man having his mother taking him wherever he goes.'

After the loss of liberty, the indeterminacy of their sentence was what the men most often mentioned as the worst feature. The anguish of the uncertainty, together with the knowledge that there was nothing they themselves could do to bring about the day of their release, was referred to repeatedly, and it was the reason why men said they would rather have gone to prison.

A variety of other matters was raised when we asked the men to say what they thought was the worst thing. Some mentioned the boredom: 'the same old thing every day . . . nothing changes'. Others talked about their fear of other patients, and their unhappiness at having to live together with people who were unpredictable or explosive, and with men who had committed terrible crimes: 'The worst thing is having to live with people you would otherwise not live with, for example, murderers'. The stigma attaching to Broadmoor was cited by some men as the worst thing: the fact that its very name struck dread into the minds of the public outside. 'There's not such a stigma about prison. When you leave Broadmoor you've never left it . . . No one will trust you again'. The idea of changing Broadmoor's name to one less charged with history and emotion was sometimes raised.

Some men said that for them the worst thing about being in Broadmoor was the distance from their families. Broadmoor is situated in rural Berkshire and its patients come from all parts of England and Wales. Inevitably therefore most are deprived of the opportunity of being visited by their families as frequently as would have been possible had they been in a local hospital. Of the mentally ill men we saw, 41 per cent had not had a visit from family or friends in the preceding three months. Almost a quarter (23 per cent) appeared to have no links at all with their families, but 42 per cent did have strong ties, and for their relations the hospital's remoteness presented serious obstacles. A middle-aged man who had come from a local hospital 8 years previously, said that for him one of the hardest things was 'not being able to see my relatives . . . they're too old to visit me now. It's 18 months since the last visit of my mother'.

As well as asking what was worst about Broadmoor, we also asked what was best. The replies could not be categorized, as too many men answered negatively: 'nothing', 'there's nothing good', 'I don't like none of it'. Those who did reply often singled out the thing that they themselves enjoyed most: for example, the football, the shop, socials, visitors. A number of men referred to the provision of asylum which was discussed earlier: 'a roof over your head, and food', 'bed, meals, bath'. One very ill man was lost for a reply when asked this question, but later sought out the interviewer again to say that the best thing about the place was that it was 'a cushy number', a view shared by a number of others. For example a schizophrenic man told us that the best thing was that: 'it's an easy life . . . you get your meals . . . and work is not pushed on to you'.

A number of men said that for them the best thing about Broadmoor had been the help they had received, either drugs or help in less tangible ways. One young man said: 'They've given me responsibility and allowed me a hand in running things . . . I help to look after very disturbed patients . . . Broadmoor has helped me to mature'.

These quotations illustrate how difficult it is to generalize about what Broadmoor meant to its psychotic patients. Although all of them were mentally ill, their backgrounds and personalities were diverse, and their reactions to Broadmoor correspondingly varied. What some men thought was the best thing seemed to others the worst. For example, some said that the best thing was the companionship: 'the laughs you get with the lads' or the fact that 'you're never alone'. To others this enforced mixing and lack of privacy was the worst thing of all. But in so far as generalization is possible, it can be said that two matters caused widespread and deep concern among the mentally ill. One was the loss of freedom, the lack of contact with the ordinary world, and the massive burden of maximum security. The other was the indeterminacy with which they were sentenced to these privations. It is the issue of discharge from that indeterminate sentence we consider in the next chapter.

4   Discharge of psychotic men

INTRODUCTION

This chapter is about the discharge decisions taken by consultants in relation to the 127 psychotic men in our sample. The great majority of these men (112) were subject to restrictions which prevented the doctors from moving them without the consent of the Home Secretary. However, the Home Secretary can only consent to a patient's move if he has adequate information about the case, and it is the Broadmoor doctors who are the source of this information and who take the initiative in drawing Home Office attention to patients they believe to be suitable for discharge. Of course the Home Secretary is then free to refuse or postpone his consent, and this is an area where difficulties sometimes arise. However, the focus of our research was the consultants' decision to recommend and not the Home Secretary's decision as to whether to accept that recommendation. There is of course considerable interest in the relationship; it might be expected that consultants would not propose discharges which they thought the Home Secretary would oppose. However there was no case in our sample where a doctor said that this had been a consideration in his decision not to recommend a psychotic man for discharge.

PATIENTS' OPINIONS ABOUT DISCHARGE

At the start of the study we had asked each man to tell us what he knew about the discharge procedure. Some patients were extremely knowledgeable about the legal issues, but all correctly identified their path to freedom as being via their consultant.

We asked men who were not due to be discharged whether they felt there was anything they could do in order to get out of Broadmoor. There was widespread recognition that it was not within their power to influence discharge decisions in any real sense. Only occasionally did a mentally ill patient think that there was anything he could do other than 'keep my nose clean'. One such man told us 'I'll have to convince them I'm O.K. . . . I'll probably have to say I wasn't hearing voices, even if I was'.

Only 57 per cent of men in the psychotic group had ever applied for discharge to a Mental Health Review Tribunal. As already noted, at the time of our study Tribunals could only recommend, not effect, the release of restricted patients. Nor were there any provisions, such as are now contained in the 1983 Act, for automatic referrals to Tribunals in cases where the patients themselves did not apply. The consensus view among the men we interviewed was that the MHRT system was, at least for restricted patients, a waste of time. It was felt that Tribunals were very unlikely to make discharge recommendations against the advice of consultants; and that even if they did, the Home Office would take the advice of the doctor rather than of the Tribunal. There was good statistical support for this opinion. In only 11 per cent of the cases where applications had been made did an MHRT recommend discharge against a consultant's advice and in none of these cases was the discharge proceeded with. Some patients, nevertheless, took the view that the Tribunal system was a worthwhile protection against being forgotten.

THE CONSULTANT'S DISCHARGE DECISION

The first question we asked the doctor at each interview session was 'Are you proposing to discharge Mr X?'. If the answer was 'yes', we interviewed him using our discharge questionnaire. If the patient was not selected for discharge we used our standard interview and also asked the doctor to complete a short schedule consisting of a number of rating scales.

Of the 127 psychotic men, 31 (24 per cent) were said by their doctors to be ready for discharge from Broadmoor. Our task then was to determine why these men were considered fit for release and why the remaining 96 men (76 per cent) were not so considered. This part of our research proved by far the most difficult to complete. We revised the interview schedules several times but were still not satisfied in the end that we always succeeded in revealing the reason for the discharge decision.

Although the decision of a consultant to recommend further detention appears on the surface to be a positive one, ('He is still ill and in need of hospital care in conditions of maximum security'), in reality this kind of decision is passive and maintains the status quo. The truly positive decision is the one which alters the status quo and recommends discharge. The question the doctor asks himself in practice is not 'Does this man need to be detained?', but rather 'What has happened to justify this man's release? Why should I now take a risk (and there always is a risk) with this man?'

REASONS FOR CONTINUED DETENTION—THE CONSULTANTS' REPLIES

Ninety-six men were not considered fit for discharge or transfer and there are 96 sets of reasons to account for such decisions. We have been obliged to reduce this abundance of information to more manageable proportions and we have attempted to do this by breaking down the population of 96 into smaller groups on the basis of the doctor's response to the question: 'What would have to happen before this man could be released from Broadmoor?'

The question was asked for 92 men, and for 56 (61 per cent) of them the answer was that their illness would have to improve. However, the prospect of this happening was judged to be remote in some of these cases, as the doctors made clear when we asked about the relative importance of time and treatment in eventually making these men fit for discharge. For 6 of the 56 men, the doctors stated that they expected neither time nor treatment to bring about any change in the man's condition; and it was evident from the histories of these men that every known treatment including electroplexy had been tried and had failed to alleviate the illness. Most of these men had diagnoses of chronic schizophrenia and had either been in-patients or had long associations with their local hospitals by the time they committed the offence which brought them into Broadmoor.

For the majority (61 per cent) of the population therefore, the main reason for detention was the lack of an effective remedy for illness. The following man exemplifies this large group. He had been in Broadmoor for 15 years, during which time he had been maintained on phenothiazine medication; he had also had four courses of ECT. His consultant reported that he had shown no response to treatment and was currently subject to delusions, hallucinations, and abnormal mood swings. The diagnosis was schizophrenia but the patient had shown no response to the drugs used to control florid schizophrenic symptoms. The function of medication in his case, and in that of others like him, would seem to have been the control of behaviour rather than the restoration of ordered thought.

In other cases, paranoid symptoms, uncomplicated by other schizophrenic manifestations, had proved completely resistant to treatment, and many years and many types of phenothiazine medication had done nothing to alter the systematized delusional beliefs of the men concerned. The illnesses from which they suffered had proved unresponsive to all treatments currently available, and represent a major challenge for psychiatric research.

For 39 per cent of the population, improvement in illness was not mentioned by the doctors as a necessary precondition for release. What distinguished these cases from the others? Table 4.1 shows details of the statistically significant differences, and it will be seen that the 'illness must improve' group was distinguished in two main ways. First, they were more often deluded and unresponsive to medication; secondly, they were much more often diagnosed as paranoid schizophrenics. For 90 per cent of the paranoid schizophrenic men the consultants gave 'illness must improve' as a prerequisite for release, but they cited this as a condition for only 59 per cent of the other schizophrenic group and for 21 per cent of the men in the third group.

Given the nature of the institution and the population, it might have been expected that the doctors would have mentioned 'improvement in mental illness' as a prerequisite of discharge for all men. The fact that it was not mentioned for as many as 39 per cent of the population was therefore of interest and caused us to examine these 36 cases in greater detail. In our schedule, the question about reasons for detention was open-ended and we derived no less than 18 categories of response when we attempted to code the data. These resolved into four main groups, not mutually exclusive, which we discuss below.

(a) Nowhere else to go

The largest single 'alternative' reason for continued detention was the lack of a suitable place to which the man could be sent. The following example illustrates the difficulty. In his early forties when interviewed, this chronic schizophrenic patient had been in Broadmoor for more than 15 years. He had made himself unpopular with his catchment area hospital because of his belligerent manner and occasional assaults against staff and other patients. He came to Broadmoor after conviction for criminal damage, though the damage done was minimal and the real reason for his admission was the refusal of his local hospital to accept him. The first attempt by Broadmoor to discharge him occurred within two years of his admission but on that and on subsequent occasions his local hospital had always refused to have him back. At the time of our study, all concerned parties were awaiting the opening of a medium secure unit in his catchment area.

The lack of a suitable alternative place did not always arise from the attitudes of the local hospital; sometimes it stemmed from the conditions which obtain in their locked wards. An example was a man in his early fifties who had been in Broadmoor for more than 14 years. Though he sometimes displayed florid psychotic symptoms, his schizophrenic illness was characterized by the negative symptoms of social withdrawal and apathy. He represented no management problem and was not thought to require the maximum security afforded by Broadmoor. An application would have been made to have him transferred had he not himself expressed a strong wish to remain in Broadmoor and had his consultant not considered that the quality of the patient's life would deteriorate were he to be moved to local hospital care. The locked wards of local hospitals are used to house the acutely disturbed. Bizarre and noisy behaviour is the order of the day there, and though most patients in such wards may recover and be returned to open and less disturbing environments, a patient transferred from Broadmoor would risk prolonged detention in such conditions.

(b) Changes in motivation or attitude

It is difficult to know what the term motivation means in the context of a continuing illness, but lack of motivation was given as the reason for the detention of some men. Perhaps 'co-operation with treatment' is a better term, for these men were regarded by their doctors as being wholly uncooperative. Certainly, in their interviews with us, a number of these patients expressed resentment at having to take medication and an unwillingness to continue with it if discharged. Others were simply very deteriorated in their social and cognitive functioning, so that they lacked insight and concern for their own welfare. Some told us they wished to remain in Broadmoor rather than be transferred to a local hospital.

(c) Insufficient time for adequate assessment

The doctors told us that a number of men were not being discharged because there had not been enough time to complete a full assessment. All such men had been in the hospital for between one and two years. This would have been considered overlong in terms of conventional psychiatric practice but the policy illustrates the specific effect of the offence committed before admission. Not only had the men concerned to be well as regards their mental state, but because of what they had done they also had to convince the doctors of the stability of their behaviour and of their willingness to continue taking drugs after discharge. Most of these men were well when we interviewed them, and their consultants told us that if they kept well and out of trouble for a further 12 to 18 months, their prospects of discharge were good.

(d) Other reasons

For a handful of men it was explicitly stated that they would have to be more open about themselves and be prepared to discuss their problems with the staff before they could be considered fit for release. The problems usually concerned sex. One such man had been a patient in Broadmoor for over 15 years and was in his early forties when we interviewed him. Diagnostic formulations had shifted from schizophrenia to personality disorder to brain damage and, finally, to an affective illness. The latter diagnosis had been made by his current consultant and he it was who had made arrangements for this man to be given direct help in coming to terms with his sexual problems and who was making his discharge conditional upon these being resolved. A disturbing feature of this case was the fact that the patient's sexual difficulties had been neglected up to this time. In a small number of other cases some particular reason was cited to account for continued detention. For example, in one case, there was an administrative problem in respect of repatriation. In another case, the patient was involved in difficult and stressful civil litigation and the consultant said he would not transfer him until a settlement was reached. His decision was based on the past vulnerability of the man to psychotic breakdown when under stress. Finally, one rather unusual case is worth mentioning. In this instance the doctor thought that the patient did not need maximum security and could safely be transferred to a local hospital. The nursing staff, however, strongly disagreed. The doctor reasoned that, as the nurses knew the patient better than he did, their view should prevail: the patient would thus not be transferred until the nurses thought it appropriate.

UNOFFICIAL REASONS FOR CONTINUED DETENTION

(a) The consultant factor

So far we have discussed the reasons for detaining patients that were given to us by their consultants. Our study of the discharge process, however, led us to conclude that some elements in it could not be elicited from our interviews with the doctors. The first of these we called the consultant factor. As would be expected in a population with an average length of stay of over eight years, nearly all the men had been in the care of more than one consultant during their stay. A change of doctor carries with it potential advantages as well as disadvantages for patients hoping to be discharged. Obviously, if a doctor has decided that a given patient is never going to be suitable for release, it is in that patient's interests to have a new consultant. On the other hand, constant changes of consultant can result in unnecessary delays in discharge, since doctors need time to get to know their patients.

The doctors tended to resist the notion that discharge was in any way dependent upon them as individuals. The facts indicated otherwise. For example, we examined the discharge pattern in one of the Houses during the three-year period when Dr X was in charge and then for the three years when Dr Y had taken over its running. The type of patient sent to the House did not change. During the three years of Dr X's period, 7 men had been recommended for transfer (an average of 2.3 per year). In contrast, during Dr Y's period 23 men were proposed for transfer, an average rate of 7.6 men each year. It seems that in those many grey areas of decision-making where 'ifs', 'buts' and 'maybes' predominate, the willingness of a doctor to review cases and make efforts to seek discharges for his patients does indeed play a role.

(b) Being 'lost in the woodwork'

Another hidden factor in continued detention was the tendency for some men to get 'lost in the woodwork'. The following case illustrates how this type of situation can arise and is also a further example of how a change of consultant can affect discharge status. The patient was a chronic schizophrenic man who had been admitted after a domestic quarrel in which he had injured a member of his family. He had never before been involved in any trouble and had been well managed by his local hospital. Once in Broadmoor, he suffered a number of depressive episodes which responded well to ECT and thereafter he proved to be 'A very co-operative patient and little problem in management'. It was, nevertheless, fourteen years before attempts were made to have him returned to his catchment area hospital. Perhaps he had been too well behaved and had thus made himself inconspicuous. In another similar case, where the patient, now in his sixties, had been in Broadmoor for many years, the Charge Nurse who was looking after him told us 'It's about time he got out . . . he's no security risk. He never asks for anything. He's been forgotten.'

It was no doubt with such patients in mind that regular compulsory Mental Health Review Tribunals were introduced in the 1983 Mental Health Act. However, procedures were in force at the time of our study which were intended to ensure that patients were not overlooked. These required the consultant to report to the DHSS every two years on the reasons why each patient needed further detention (section 43 of the 1959 Act, and Paragraph 166 of the Ministry of Health Memorandum on the Act, 1960). The object was to ensure that the case of every patient would be systematically and thoroughly reviewed, but in practice it did not work that way. The DHSS, to whom the reports were sent, very rarely commented on them; and in the hospital the reports came to be treated as paper formalities, completed in standard phrases such as 'he is deluded, insightless, and in need of care.' It was thus rare for these reports to occasion a fresh appraisal of a patient's need to be detained: this was much more likely to follow from the regular case conferences which some, but not all, consultants were undertaking at the time of our study. The 1983 Act now requires consultants to send annual reports on restricted patients to the Home Office (section 41(6)), and it will be interesting to see the effect of this new provision.

MEN CONSIDERED SUITABLE FOR DISCHARGE

Thirty-one men were said by their doctors to be suitable for discharge. For three men repatriation was proposed and one man was considered ready for discharge to a hostel. For the rest of the group (27 men), transfer to other hospitals was envisaged. Ten men were proposed for medium secure units and 16 for ordinary psychiatric hospital wards. One man was proposed for the Eastdale Unit at Balderton Hospital, a centre which caters for former special hospital patients.

Having established where he was to go on discharge, we next attempted to find out what factors had brought the man to the doctor's attention with regard to discharge – not an easy question for doctors to answer. On the basis of pilot work we had carried out on our interview schedule, we produced 11 possible factors, which were not mutually exclusive. As it turned out, only one of these factors emerged as being numerically significant: almost half the men (48 per cent) had come to the doctor's attention because of a case conference.

More than anything else, the responses of the doctors emphasized the importance of routine and thorough case reviews.

Discharge decisions represent the end result of a complex procedure involving many factors. During the interview, each doctor was asked to complete a checklist of variables which, again on the basis of our pilot work, we believed might be important in making discharge judgements. To the items on this list, the doctors were asked to add any others which had played a role in leading them to recommend the discharge of the man about whom they were being interviewed. Each item on the checklist was provided with four alternative responses ranging from 'not at all important' to 'very important'. These responses were weighted so that 'very important' was given a value of three and at the other end of the scale 'not at all important' was given a value of nought. Total scores were derived for each factor and are presented in Table 4.2.

Top of the list was a change in the man's mental state and this variable correlated with change in behaviour ($r = 0.52$) and with likelihood of reoffending ($r = 0.54$). Indeed, these three categories were the only ones rated as important for a majority of men in the group and reflect the view of the doctors that, more than anything else, what determined discharge was a change of mental state. A principal component factor analysis of the relationships between these elements confirmed the importance given to mental state and behaviour by the medical staff. Most variables loaded highly on this first component, the exception being the administrative element of whether or not a new facility was now available for the man in his local hospital catchment area.

When a doctor told us that a patient's mental illness had improved, we asked for information about when the improvement had occurred. As a rule we found this information impossible to obtain. Very few of the doctors we interviewed had personal knowledge of what their patients were like when they were admitted to Broadmoor, and, apart from the admission assessment reports, the patients' case notes rarely contained detailed information about mental state.

The importance of delusional symptoms in the relationship between violence and psychotic illness is well established (Walker & McCabe, 1973) and has been demonstrated again recently by Taylor and Gunn (1984). It was apparent that reduction in the frequency or intensity of such symptoms plays a major part in transfer from Broadmoor, but most doctors were unable to tell us when such changes had taken place. We could only conclude from this, once again, that the regular and thorough review of patients, appropriately recorded, was very important in the matter of discharge.

A few of the men being discharged did not appear to have changed in relation to their illness. In one case the patient was considered to have been made adequately risk-free by virtue of increasing age and physical debility. Another patient presented an unusual story. He had been recommended for transfer by his consultant some years before our research began, but his catchment area hospital refused to readmit him and no further action was taken by the Broadmoor authorities. The patient then took it upon himself to approach a variety of other hospitals and was eventually offered a place. His action illustrates how extraneous circumstances, unrelated to mental illness, can affect the discharge process.

WHO NEEDS BROADMOOR'S MAXIMUM SECURITY?

A basic question in our research was: how many men remain in Broadmoor for any reason other than their perceived need of maximum security? A saying popular in the hospital and attributed to one of its earlier Medical Directors, is that half the population could be discharged, the trouble was in knowing which half. For each man not on the discharge list we asked the doctor whether Broadmoor's maximum security was required, or whether the patient's security needs could be met in less stringent conditions.

The answers showed that nowadays the doctors in Broadmoor do feel that they know which of their patients require its maximum security facilities, and it is less than half. Of the 123 men in our mentally ill sample for whom information was available, only 35 (28 per cent) were considered to need maximum security; 31 men (25 per cent) were currently being processed for transfer, another 45 (37 per cent) were said to be suitable for conditions of less security, and 12 patients (10 per cent) were described as probably suitable.

When patients not on the discharge list were described as being suitable for less than maximum security, we asked the consultants why transfers to medium secure units had not been attempted. The replies were nearly always the same, the doctors believed such units would not accept patients who needed long-term care. In only one case was the lack of a local secure unit mentioned.

WHAT VARIABLES WERE ASSOCIATED WITH DISCHARGE STATUS?

Since the group of men not being discharged contained a large proportion considered by their doctors to be ready for less secure establishments, we divided our population into three categories, to determine which variables were associated with discharge status. The first group was made up of men who were on the discharge list, the second group comprised those who were not being discharged but were not thought to require maximum security, and the third group contained the men said to require the degree of security provided by Broadmoor. These groups did not differ in terms of their age, age at admission to Broadmoor, or the length of time they had spent there. They were also similar in regard to previous experience of prison and psychiatric in-patient care.

Table 4.3 contains a list of those variables which distinguished between the three groups in a linear fashion. There were six such variables and it is worth pointing out that none of them related to the offences for which the men had been hospitalized. It might have been expected that the severity of their offences would be reflected in the consultants' assessment of the men's need for security, but this was not the case.

When patients are admitted to Broadmoor they are asked to complete psychological tests relating to intelligence and personality, the Minnesota Multiphasic Personality Inventory (MMPI) (Hathaway & McKinley, 1943) being used to assess personality. The results were recorded by us in our data collection. It will be seen from Table 4.3 that four of the differences between our groups related to MMPI subtests. In each case, high (pathological) scores were associated with non-discharge status and the need for maximum security. Some care must be exercised in discussing these results as a significant proportion of these men (30 per cent) did not complete MMPI forms on admission. The most relevant and statistically significant scales concern the reporting of Anxiety (Welsh, 1956) and Paranoid tendencies (Baggaley & Riedel, 1966). It should be noted that the latter is not the clinical 'Paranoia' scale but a much narrower scale derived from factor analysis of the MMPI. The scale is designed to measure suspiciousness and the tendency to attribute to others harmful intentions and behaviour. The finding in regard to this scale is especially interesting because it reflects an attitude or trait present at time of admission. It may be that the type of paranoid disorder resistant to modification and control by phenothiazine medication is related to personality characteristics or traits rather than symptoms (cf. Foulds, 1965), and that within the schizophrenic population a distinction should be drawn between symptom- and trait-related paranoid ideation.

The lower part of Table 4.3 shows which of the consultants' assessments distinguished between the groups in a linear fashion.

Unimproved mental state and the persistence of delusional symptoms were clearly adverse factors in the discharge process. However, it is worth noting that delusional symptoms did not exclude men from the discharge list: a quarter of patients being discharged were described as deluded.

Our next step was to see what factors distinguished the men deemed to need maximum security from patients in the other two groups. The criterion was that of need for security and Table 4.4 shows which variables were found to be associated with this. Of major importance here was the man's diagnostic status: patients diagnosed as having paranoid schizophrenia were most prominent within the group needing the highest security. Another factor distinguishing the men needing maximum security was that a higher proportion of them had been known to their doctors for less than two years.

We next examined the data to see which variables distinguished men in the discharge group from the others. Details of these items are presented in Table 4.5. The most statistically significant item was the consultants' estimates of control achieved by medication ($p < 0.001$). The discharge list included the majority of patients for whom such control was rated as very good and none of those for whom it was rated as inadequate.

One of the questions we asked the doctors was whether they were aware of anyone outside the hospital, such as a relative or friend, who was pressing for the man to be discharged or to be kept in hospital. Six men were said to have had people pressing for them to be kept in and of these six, one was on the discharge list. Twenty-one men were said to have someone pressing for their discharge and ten of them were in the discharge group: this item was one of the most statistically significant discriminators between the groups. The relationship must be interpreted cautiously. It does not mean that such pressure was solely or even mainly responsible for the discharge of these men, but rather that, other things being equal, pressure or interest from outside conferred an advantage in the matter of discharge. The fact that this interest did not confer benefit on all patients was very evident in our interviews with the doctors, who were quick to point out that no amount of outside pressure would persuade them to recommend the release of some men. In general, the doctors maintained that outside pressure was of negligible importance in their decision-making. The statistical evidence does not support this view, and the discrepancy may be due to the fact that the expectation of support from someone outside the hospital tipped the balance in favour of making a discharge decision, when other factors were equal. Certainly this type of explanation would account for the group difference and at the same time be consistent with the doctors' view of the relative unimportance of outside pressure.

In its investigation of mentally abnormal offenders, the Butler Committee (Home Office, DHSS, 1975) reported that many of its witnesses had testified to the usefulness of restriction orders in so far as they allowed for the conditional discharge of patients. Tidmarsh (1981) has pointed out that 'unrestricted patients may find their stay in maximum security unnecessarily prolonged because responsible medical officers . . . lack the confidence to discharge patients whose insight is often the first casualty of their mental disorder'. It was apparent from our interviews that other Broadmoor consultants were of the same opinion, conscious as they were that it was only by means of a conditional discharge that they or the Home Office could exercise any control over the discharged patient. Our data reflected these views. There was a trend ($p < 0.09$) for the discharge group to include a higher proportion (93 per cent) of restricted patients than did the other two groups combined (79 per cent). Though not playing a primary role in the discharge process, this trend seems to reflect a bias operating against men who could only be given absolute discharges if freed. It was among the group not held to require maximum security but not being discharged, that the highest proportion (22 per cent) of non-restricted patients was found.

Finally, we examined the data to see what characterized the men described as suitable for less security but not for discharge. Table 4.6 contains a list of the variables which distinguished this subgroup from the others. Two features are apparent. The first is the disproportionately high number of men in the group who were receiving more than one major tranquillizer. The second element relates to the admission offence: fewer of these men had caused serious injury and more of them had offended against strangers. The picture suggests that the group contained a disproportionate number of men who were debilitated by their chronic illnesses and who had gravitated to Broadmoor via the local hospital system because of their assaultativeness.

THE APPROACH TO MOVING PATIENTS OUT

As we have seen, many different factors play a part in the consultant's decision to recommend transfer. The overriding principle which affects the way in which these factors are regarded is that of caution. The doctors are dealing with men who have (in the main) caused serious injury to others and nothing is more important to them than to ensure, as far as is possible, that such events do not recur. Proposing patients for discharge introduces uncertainties and possible risks, and there is thus inevitably an inherent bias for the doctors to maintain the status quo and to keep their patients in Broadmoor. Additional factors act to reinforce the bias. Consultants know that so long as their patients are in Broadmoor they are not only secure but well cared for. Indeed, as we have seen, for some psychotic patients it is their view that the standard of care in Broadmoor is superior to that available elsewhere. Thus, even if transfer to a local hospital seems a possibility, some measure of further detention in Broadmoor is not usually seen as particularly harmful. On the contrary, it means that there is more time to ensure that the patient's condition has really stabilized.

There are thus strong factors tending to militate against the making of transfer recommendations. A concomitant of this situation was something that was striking to outsiders like ourselves: the release from Broadmoor of a patient regarded as ready for a move was not generally seen as a matter of urgency. The unnecessary detention of a person in a maximum security hospital can be seen as an infringement of civil liberties, but our examination of patients' records and our interviews with their doctors did not lead us to think that consultants usually saw it in this way. The following case illustrates how little sense of urgency might accompany a consultant's decision that his patient no longer needed to be in Broadmoor. The doctor had noted in the records that the patient had been well behaved for two years and was suitable for an open hospital. No action followed. It was not until there happened to be a change of consultant more than two years later that steps were taken to find an NHS bed.

DISCUSSION

The illnesses from which most of the men in Broadmoor suffer are chronic. Some remit from time to time, some are under good control by medication, but in practically every case the men who formed this population must be considered vulnerable to their illnesses for the rest of their lives. An appreciation of this fact is fundamental to any understanding of the nature of decision-making about their removal from the special hospital system.

Probably the most important point to emerge from our inquiries is that only 28 per cent of the mentally ill men were thought by their doctors to need the maximum security that Broadmoor provides.

About a third of the resident mentally ill male population in the hospital was included in our study. It is clear that were security considerations alone to determine detention in Broadmoor its population would fall dramatically.

We asked the consultants why they were continuing to detain so many men whom they thought suitable for lesser security. They said that they did not think Medium Secure Units would be prepared to take patients who needed long-term care and that they were often dubious about the quality of care and supervision available in ordinary local wards. Given that so many men judged suitable for less security were not being proposed for transfer, it seemed to us regrettable that Broadmoor provides no gradations of security. All the psychotic men were subjected to the massive restrictions on their liberty which the doctors thought only 28 per cent of them needed.

5 Discussion: psychotic men

RESEARCH

Before discussing our research findings, there is one point that seems
to us important to make. Broadmoor houses, as our study showed, a
long-stay population of mentally ill patients presenting with a vast
range of, as yet, incurable psychopathology. The hospital thus offers
excellent opportunities for carrying out fundamental research into
the nature of schizophrenic illnesses. The effort which goes into
establishing the diagnosis of patients arriving in Broadmoor is very
impressive, but once the assessment period is passed, there seems to
be little systematic examination of patients' mental states. An incen-
tive to better recording might come from research, and if the hospital
were to obtain or gain access to the latest technological advances in
monitoring brain function, it could well become the focus of major
research in this important area.

TRANSFERS FROM PRISON

If a prisoner becomes ill and is transferred to a special hospital, he is
liable to be detained there beyond his earliest date of release from
prison. Grounds (in press) has shown that most transferred prisoners
are in fact detained after their sentences have expired and this was
true of such men in our sample. It seems to us unfair, and the men
concerned were of the same opinion, that a person who happens to
be serving a determinate sentence when he falls ill should thereby
become liable to indefinite detention in maximum security. Were the
same man to become ill at home, he would not be sent to a special
hospital but to his local hospital. We think it would be fairer if
hospital orders expired at the prisoner's earliest date of release and
that, if at all possible, patients should be moved out of special
hospitals by that date. If further detention in maximum security were
thought necessary, then we suggest that this should be arranged
under civil procedures involving outside doctors and social workers.
We would also propose as a further safeguard that there should be
reference to a Tribunal within a month of any civil treatment order
being made.

ISSUES RELATING TO DETENTION AND TREATMENT

As regards the psychiatric treatment of mental illness, there is nothing special about Broadmoor; no methods are used there that are not available in local hospitals. What is special, is its ample nursing staff/patient ratio (Hamilton, 1985) and the secure containment offered by its locks and high wall.

When we asked the doctors about the reasons why their mentally ill patients were being held, reference was made in the majority of cases to the need to achieve better control of the patient's illness. It became apparent, however, that the criteria for continued detention in Broadmoor were very different from the detention criteria used in ordinary psychiatric practice. In local hospitals, if a patient needs to be detained against his will, the period of that detention is determined by his mental state. Once restored to better health and behaviour, he is discharged from the detention order, although he may continue to live in the hospital as an informal patient. In Broadmoor, different criteria apply before a man is even suggested for detention in another hospital, for it is not only his present state but his past behaviour that is taken into account. In these circumstances it becomes difficult, certainly for the patient, to separate penal from psychiatric custody.

For most patients in our sample, admission to Broadmoor followed directly from the commission of a serious offence; they were sentenced to the hospital by a court. That court, having decided on the one hand that a penal disposal would not be suitable, nevertheless imposed Home Office restrictions on the patient's discharge, thereby taking this out of the control of the doctor in charge of his care. In this way, the restricted order patient is set apart from every other class of detained psychiatric patient.

An important aspect of the restriction order is that it ensures that progress of the discharged patient will be supervised after release and that he can be recalled to hospital if necessary. Many of the Butler Committee's witnesses thought it valuable for these reasons. (Home Office, DHSS, 1975, para 4.17). As we saw in the previous chapter, Broadmoor doctors were more prepared to propose the discharge of restricted patients whose progress is subject to such monitoring, than of non-restricted men whose discharge has to be absolute. Given the value of conditional discharges with the provision for supervision and recall, it seemed to us unfortunate that it could only be used for patients subject to Home Office restrictions. There are arguments for introducing a form of hospital order which would allow doctors to move and discharge patients without reference to Home Office authority, but which would nevertheless provide powers of conditional discharge.

Broadmoor is unusual in caring for both mentally ill and non-ill patients. As will be seen, some of the latter told us that they found it hard, as sane, young, intelligent men, to be in a community consisting in the main of older psychotic patients. It seemed to us however that the mentally ill also suffered disadvantages by being confined alongside non-ill offenders. The massive security that obtains in a hospital detaining active young men categorized as psychopaths is not necessary for many of the chronically debilitated schizophrenic patients, and adversely affects their living conditions.

Broadmoor doctors sometimes welcome the admission of non-psychotic offenders on the grounds that they are interesting; the prospect of caring for an entire population of schizophrenic patients is not attractive. There is however a danger that in this situation the interests of the psychotic patient may be neglected or pushed into second place, if only because the non-psychotic are so much more active and alert. One of the nurses made this point to us when we were discussing the discharge prospects of a mentally ill man: 'He's been forgotten . . . it's the psychopaths that get all the attention'. From our study of the discharge process it certainly seemed that psychotic men were more likely than others to have disappeared into the woodwork and been overlooked. As part of our coding procedure, we noted whether men were of special interest in regard to various factors, including 'being lost in the woodwork' and we found that this variable had been recorded twice as often for psychotic as for non-psychotic men ($p < 0.08$).

SECURITY NEEDS

Perhaps the most striking finding of the research was what the doctors said to us about the patients' need for maximum security. Of the psychotic men, 25 per cent were being processed for discharge, 47 per cent were said not to need maximum security (including 10 per cent described as probably not needing it) and only 28 per cent were said to require Broadmoor's maximum security.

There are obviously profound implications in the finding that so massive a proportion of Broadmoor's mentally ill offenders are judged by their doctors not to need maximum security. The DHSS itself has pointed out that in the special hospitals: 'The regimes of care and observation are such that they can only be justified when the highest level of security is required and no lesser degree of security would provide a reasonable safeguard to the public'. (DHSS, 1983, para 266).

Why had the consultants not taken steps to secure the transfer of the men who were not on the discharge list, yet were described as being suitable for regional secure units or lesser security? The explanation most commonly given by the doctors was that these patients would not be accepted locally, and that regional secure units would not take men who needed long-term care. It seemed to us that reliance should not be placed on assumptions, but that discharge or transfer should always be sought, once a patient is no longer considered to need maximum security. Apart from anything else, the local health and hostel authorities cannot otherwise know what provision needs to be made for these patients.

In some cases, transfer proposals were not made for psychotic men because doctors believed they were better off in Broadmoor than elsewhere. It was held that seriously ill long-term patients enjoyed in Broadmoor a better standard of nursing and a more agreeable ward environment than was available for them in local hospitals. The lack of good long-term care in the National Health Service was a recurring theme raised by the consultants. If Broadmoor staff believe that by transferring their patients they are relegating them to inferior, inadequately staffed hospitals, this is obviously a major disincentive to the making of transfer proposals.

If security needs alone then were to determine the size of Broadmoor's mentally ill population, it would fall dramatically. However, this seems unlikely to happen given the shortage of alternative long-term facilities. Regional secure units do not provide long-stay beds, and with the closure of the old psychiatric hospitals, such beds have become increasingly scarce in the local National Health Service system. In these circumstances it would clearly be valuable if some graduated system of security were available within Broadmoor; it would ensure that patients were not subjected to greater restrictions than were necessary. It would also, where appropriate, enable patients to get used to living in more normal and less restricted conditions, and this would help eventually with their transfer or discharge: local hospitals and hostels are understandably nervous of accepting people who have for many years lived under a regime of maximum security and control, and about whose response to conditions of less security little or nothing is known.

There is within Broadmoor an inevitable tension between security and treatment needs. It is styled a secure hospital, but as we have seen, many of its patients also view it as a prison. Perhaps the conflicting demands of care and security can never be wholly resolved.

Ideally, we think that smaller, regionally based facilities would best meet the needs of those who are said to require long-term care but not maximum security. They would ameliorate the problem of distance from family and friends and make it difficult for patients to be overlooked. It is true that such facilities could not provide the range of provision available within a large hospital such as Broadmoor, but by being more flexible and relaxed in their approach to security they could enhance the quality of patients' lives.

Part II: Psychopathic disorder

6 Men admitted with a legal classification of psychopathic disorder

LEGAL ADMISSION PROCEDURES

The entire population of men resident in Broadmoor under the legal category of Psychopathic Disorder (PD) in April 1982 forms the group considered in this chapter. There were 117 such men, and when the population was collected, they constituted about one-quarter of the hospital's resident male population. The average length of stay of the 117 men was eight years (s.d. 4.4 years, range 1–21 years).

The legal procedures under which patients are admitted to Broadmoor were described in Chapter 2. Of the 117 men in the sample, most (102 men, 87 per cent) had come directly from courts under restricted hospital orders. All but four of these orders had been made without limit of time. Twelve men (10 per cent of the population) had been transferred from prison after sentence, and when the data were collected, eight of them had passed their latest date of release and were no longer subject to restrictions. Three men were admitted under hospital orders without restriction on discharge.

THE ADMISSION OFFENCE

Details of the admission offence are listed in Table 6.1. Almost a third of the population had committed homicide, and 73 per cent had been convicted of offences carrying a maximum sentence of life imprisonment.

Offences of violence were mainly against women: the victims were female in 64 per cent of cases, male and female in 8 per cent, and in only 27 per cent of cases was the sex of the victim male. The bias is a reflection of the sexual nature of many of the offences. Irrespective of the offence for which they were convicted, almost half the men (48 per cent) had committed or attempted a sexual assault. (The finding contrasts with the picture for the mentally ill population, where only 12 per cent of offences involved a sexual component.) Overall, 27 per cent of victims had been children aged 13 or younger, and other teenagers accounted for a further 14 per cent of the victim population. People aged 60 or over constituted 11 per cent of the victim group.

In only 15 per cent of cases involving violence were close relations involved; no psychiatric hospital staff were victims and the largest victim category by far (61 per cent) was that of stranger. The contrast with the mentally ill men is striking, comparable figures for that group being 35 per cent for close relatives, 12 per cent for hospital staff, and 27 per cent for strangers.

There is no such thing as a typical offence or case, but the following two summaries are presented in order to indicate something of the variety of men and circumstance represented within this population.

The first man was in his early twenties when admitted to Broadmoor and had been there for more than five years when we interviewed him. In the course of a burglary, he had attacked the householder and there had been a sexual component to the assault. The sexual attack and the fact that the female victim was elderly were regarded as evidence of psychological disorder and resulted in his being admitted to Broadmoor. In character, he was a rather impulsive young man of limited intelligence whose home background had been chaotic. He had been in trouble from an early age for minor delinquency, but until his admission offence had never been convicted of a violent crime.

The second man was in his early sixties when we interviewed him. He had been in Broadmoor for more than ten years. Long-standing interest in a dangerous form of sexual activity with children had resulted in the death of one of his victims. His aberrant sexuality apart, his life-style had been conventional and unremarkable. In character, he was a very anxious man. In his admission assessment, the consultant concluded: 'the therapeutic prospects are gloomy and I suspect he will need the long-term asylum function of Broadmoor'. The prophecy has been fulfilled.

Both men were admitted to Broadmoor having been described as psychopathically disordered but it is difficult to determine what they had in common other than that they had both committed a lethal or almost lethal offence which had a sexual component to it.

The seriousness or potential dangerousness of an offence cannot always be judged by the nature of the conviction. In order to overcome this problem, we read the descriptions of the circumstances surrounding each offence and noted whether its gravity appeared to account for the admission to Broadmoor. Seventeen (15 per cent) of the men admitted under the category of psychopathic disorder seemed to us to be of special interest because their offences did not really account for their admission to a maximum security establishment.

One such man had come to Broadmoor in his late teens and had been there five years when we saw him. His admission offences were: (1) breaking and entering (though not for gain) and (2) criminal damage. It was the latter offence which was considered pathological, although it was not dangerous and could have been considered, had the man wished to present it as such, as the impulsive act of an immature youth. Instead, in his interviews with the psychiatrists who saw him on remand, he gave an account of long-standing fantasies of sexual and other sadistic acts. He was duly recommended for Broadmoor. When we asked him about his behaviour at the time of his admission he said that he had exaggerated his 'symptoms' in order to get into a secure hospital, as he felt that he would only have walked out of a local hospital had he been sent to one.

If an offender is not categorized as mentally ill, and if the offence of itself does not place him within the obviously dangerous category, what criteria are used to decide that he requires detention in a special hospital? Usually at the heart of these cases was the doctors' belief that the man was liable to behave dangerously in the future, even though his current offence was not particularly serious. Various factors inclined the doctors to this belief. In one case, it was the fact that the man had on a previous occasion committed a serious sexual crime. Frequently, however, it was the men's own confessions that led the consultants to feel serious concern about future dangerousness: in over a third of these cases, the offenders had confessed to long-standing and potentially dangerous fantasies (usually sexual) for which they requested help. The detention of offenders for what they might do rather than for what they have done is clearly a serious matter. As we shall see later, some of these men created difficult problems when their possible discharge came to be considered.

As explained in Chapter 1, when we interviewed the Broadmoor doctors, we asked them to provide a diagnosis for each man. There was an interesting relationship between the specificity of the diagnosis with which we were provided and the offence at admission. Of those men who had killed someone, 88 per cent were given a diagnosis that was no more specific than 'psychopathic disorder' or 'unspecified personality disorder'. Such diagnoses were provided for only 61 per cent of men with less serious types of admission offence ($x^2 = 8.06$ 1 df $p < 0.01$). It would seem that there was an inverse relationship between the specificity, and perhaps validity, of a diagnosis and the gravity of the crime at admission.

PREVIOUS CONVICTIONS

The great majority of men (85 per cent) had a criminal record prior to the admission offence but the extent of their previous involvement with crime was wide-ranging. Just over half had been in either prison or borstal on at least one occasion before coming to Broadmoor, and 30 per cent of men had served more than one sentence. Details of the types of crime in which they had been involved are presented in Table 6.1. Larceny was the only conviction that had been incurred by a majority of the men. A substantial proportion (43 per cent) had no previous convictions for dangerous offences (i.e. violence, sexual or otherwise, or arson).

MEN ADMITTED FROM PRISON AFTER SENTENCE

Nine of the prison transfer group had been serving determinate sentences at the time of transfer, and three were serving life sentences. As noted already, if a prisoner is serving a finite sentence, the effect of a transfer to a special hospital can be to prolong detention beyond his release date until either his doctor or a Mental Health Review Tribunal agrees to his discharge. Eight of the nine men so transferred in the present group had been detained for longer than they would have been had they remained in prison. All had passed their latest date of release and were no longer subject to Home Office restrictions.

In three cases, the transfer from prison appeared to have been made in order to prevent the man's discharge into the community. The cases raised important issues, illustrated by the following example. The man concerned was sent to Broadmoor a couple of weeks before his earliest date of release, having been imprisoned for a sexual assault. He had been assessed for a psychiatric disposal at the time of his conviction, but had made it clear that he was unwilling to co-operate with treatment. He was therefore sentenced to imprisonment. The prison authorities believed he was likely to commit further serious offences, and suggested a transfer to Broadmoor at the end of his sentence. In other words, a Broadmoor bed was sought so as to enable the Home Secretary to issue a transfer order, to circumvent the court's determinate sentence, and to allow the offender to be indefinitely detained.

In all three cases, the prison doctors who initiated the transfer and the Broadmoor doctors who supported it did not have more information about the prisoners' likely future behaviour than was available to the courts at the time of sentencing. It was not the men's behaviour in prison but their previous history of offending that led the doctors to recommend, and the Home Secretary to direct, transfer to Broadmoor. The wish to prevent further violent offences motivated them: the question of treatment was virtually an irrelevance, for it was recognized by Broadmoor's consultants that a prisoner who is not mentally ill and is transferred just as his release approaches, will be hostile, bitter and uncooperative.

Cases like these were not numerous (three in our sample of 117 residents), but they still occur, as we saw at the weekly admission meetings, and raise difficult issues. In one of the cases discussed at a meeting we attended, a prisoner was coming to the end of a sentence for a planned stabbing. He had a deep resentment against the victim, and made it clear that he intended to murder him when he was free. The prison doctors asked a Broadmoor consultant to visit, with a view to a possible transfer under the Mental Health Act. The consultant reported to his Broadmoor colleagues that he found the prisoner to be 'perfectly normal' in psychiatric terms, but determined to kill.

The Broadmoor doctors were divided in their views about what action to take. Some thought that the overriding consideration should be the prevention of a murder attempt. Although the prisoner was described as perfectly normal, and nothing new about him had emerged since he was sentenced, it was argued that his aggressive behaviour would permit him to be hospitalized under the category of psychopathic disorder. Other consultants felt that this was stretching the Act too far, since there was no evidence of treatability, and none of mental disorder other than the offending behaviour. 'It's not our job to mop up judicial mistakes . . . and to take people only for custody'.

PSYCHIATRIC FACTORS

(a) The concept of psychopathic disorder

In considering the term 'psychopathic disorder', a distinction must be made between the legal use of the term, the concept as used in psychiatry, and the concept used by psychologists. Our report deals with the first of these, the use of the term in law. However, something needs to be said about the historical development of the concept of psychopathic disorder in psychiatric nosology.

Lewis (1974) in his seminal paper, 'Psychopathic Disorder: a Most Elusive Category', provides an outline of the historical development of the term. He quotes Pinel as having in 1801 used the phrase *manie sans délire* which he defined as lack of emotional control and sometimes impulsive violence in the setting of intact cognition. Lewis points out that it was not until 1835 that Pritchard in 'A treatise on insanity and other diseases affecting the mind' first used the term moral insanity. Pritchard accepted Pinel's view, but enlarged the concept to mean '. . a morbid perversion of the natural feelings, affections, inclinations, temper, habits, moral dispositions and natural impulses, without any remarkable disorder or defect of the intellect or knowing and reasoning faculties and particularly without any insane illusion or hallucination'. It should be noted that, despite changes in the wording, this two-pronged definition of psychopathic disorder remains with us; i.e. it is defined as bad or socially unaccept-able behaviour without overt intellectual defect or symptoms of mental illness to explain that behaviour.

The next major contribution traced by Lewis in regard to English psychiatry came in 1874 from Maudsley in 'Responsibility and Mental Disease'. In this, as quoted by Lewis, Maudsley stated 'as there are persons who cannot distinguish certain colours . . . so there are some who are congenitally deprived of moral sense'. He further went on 'moral insanity is a form of mental alienation which has so much the look of vice or crime that many persons regard it as an un-founded medical invention . . . (the individual) has no capacity for true moral feelings; all his impulses and desires to which he yields without check are egotistic . . . his affective nature is profoundly deranged, and its affinities are for such evil gratifications as must lead to further degeneration and finally render him a diseased element which must either be got rid of out of the social organisation or be sequestered or made harmless in it'. As we shall see later, the absolute nature of this definition of moral insanity recurs again in the description employed by Cleckley (1976).

Meanwhile, continues Lewis, on the continent of Europe, Magnan, developing an idea of degenerative processes introduced first by Morel in 1839, described four subgroups within the category of *héréditaires dégénerés*. These were, idiots, imbeciles, the feeble-minded, and *dégénerés supérieurs*, this latter term being reserved for those people who had average or superior intellects but moral defects. The important point to note here is the link which this established with mental handicap. Lewis states that the term introduced by Pritchard (moral insanity) continued to be used in English-speaking countries instead of psychopathy until the 1920s and he traces this, in part, to the use of the term 'moral imbecile' in the 1913 Mental Deficiency Art and the term 'moral defective' which had been used in the 1927 Amendment Act. The term psychopathic disorder was not given any legal recognition until the advent of the 1959 Mental Health Act.

According to Lewis, the word psychopathic was first introduced by Koch in Germany in 1891. He had employed the term 'psychopathic inferiorities' as a plural term to include 'all mental irregularities . . . which influence a man and cause him . . . to seem . . . not fully in posession of normal mental capacity, though even in the bad cases the irregularities do not amount to mental disorder'. Koch presumed that such disorders had a physical basis but admitted that this could not be demonstrated. Over the years, the use of the word inferior was dropped and the term psychopathic gradually took over from moral insanity in describing the disordered who could not be labelled ill or intellectually handicapped.

Lewis uses the various editions of Kraepelin's textbook to demonstrate the changing nature of the concept and concludes that 'successive editions saw him struggling with little success to cope with the task of shaping categories out of the rich variety of human character and conduct'. In this connection, Lewis quotes Allport; 'All typologies place boundaries where boundaries do not belong. They are artificial categories . . . each theorist slices nature in any way he chooses and finds only his own cuttings worthy of admiration'. Having considered all attempts at definition, Lewis states (p. 139) 'the conclusion of the whole matter is somewhat gloomy'.

A notable omission from Lewis's review is any reference to the writings and opinions of Harvey Cleckley ('The Mask of Sanity', 1941 through five editions to 1976). Though gaining wider acceptance in North America than in Europe, Cleckley's concept has nonetheless been influential in this country and has been the starting point for the work of psychologists such as Hare in Canada (cf. Hare & Schalling 1978). A useful summary of Cleckley's ideas is presented in Prins (1980), though Cleckley's books are to be recommended as they provide a welcome relief from the narrowness of most psychiatric texts.

Cleckley describes the condition of psychopathic disorder as a kind of moral psychosis in which the individual has no sense of the boundaries which provide us with our concept of self. The people he describes in his book as examples of psychopaths are almost all of well above average intelligence and have very superior social skills, at least at the superficial level. Most could be described as chameleons in regard to their personality, suiting their conversation and behaviour to what they consider others wish it to be, rather than with reference to any internalized structure. In the latest edition of his book Cleckley proposes that there may be a neurological basis for the type of disorder he describes, a sort of semantic aphasia which prevents the psychopath from learning the emotional significance or meaning of words, and their significance for others. He speculates that this type of function would be located in the dominant temporal lobe. There is a strong similarity between this type of explanation of psychopathy and that proposed by Koch at the end of the nineteenth century. Also, as already noted, Cleckley's definition bears a striking resemblance to that of Maudsley's one hundred years earlier, in that both proposed that the condition was represented by an absolute lack within each individual of a faculty present within the non-psychopathic.

The great strength of Cleckley's work is that he does not restrict himself to criminal or incarcerated populations but describes people from all walks of life. For example he discusses the psychopath as businessman or professional man. As with other descriptions, the concept as described by Cleckley is strong on face validity; it uses terms which are familiar and carry meaning. It is weakest however when examined for its incremental validity. What is the specific psychiatric and medical component to the so-called mental disorder? Is the term anything other than a doctor's way of saying that which it is the capacity of anyone to discern and describe, so that the terminology is all, and without it, the disease 'psychopathy' does not exist as a medical entity?

A major contribution to the legal definition came from Sir David Henderson (Henderson, 1939). His description of the psychopath excluded those who were mentally defective and incorporated the notion of lifelong antisocial behaviour which was difficult to influence 'by methods of social, penal and medical care and treatment . . . The inadequacy or deviation or failure to adjust to ordinary social life is not mere wilfulness or badness which can be threatened or thrashed out of the individual but constitutes a true illness for which we have no specific explanation'. There are traces here of Maudsley's view and the condition is defined negatively from the psychiatric standpoint. In essence, we are told that persistent incorrigible bad behaviour which cannot be controlled by normal means represents an illness. Henderson subdivided the psychopathic population into three groups, aggressive, inadequate, and creative, but, as the Butler Report (Home Office, DHSS, 1975 para 5.10) points out, these subdivisions 'were brought together in one category on a highly complex psychological theory and were not based on descriptions of observed behaviour'.

Henderson's classification was used by Hill and Watterson (1942) in a paper entitled 'EEG studies of psychopathic personalities'. Among their 66 aggressive psychopaths they found that 43 men (65 per cent) had EEG abnormalities as did 12 (32 per cent) of the 38 inadequate psychopaths and 17 (50 per cent) of men who had psychiatric problems such as epilepsy and schizophrenia in addition to psychopathy. The authors concluded 'We have little doubt that an abnormal EEG constitutes for its possessor a handicap in the business of biological adaptation, failure of which may show itself, as in our present series, in undesirable social behaviour'. They favoured the idea that an abnormal EEG might reflect cortical immaturity, the behavioural correlate being relative lack of control. In fact, the results presented in the paper shed little light on the validity of the concept of psychopathy. Of the 104 men described as psychopaths, without other symptoms, only half (53 per cent) evidenced EEG abnormalities and the other half did not. The bias was for abnormality to be noted most often within the aggressive group of men, and the study could have been reported with greater accuracy had reference been made more simply to the relationship between aggressive behaviour and EEG findings. The concept of psychopathy served largely to confuse the main result.

Hill and Watterson's paper is quoted here at length because it, along with similar reports, influenced the deliberations of the Royal Commission (1957), whose report to Parliament formed the basis for the 1959 Mental Health Act, including of course the introduction of the term psychopathic disorder. These studies in the 1940s and early 1950s had encouraged the hope that it would soon be possible to make the diagnosis and treatment of psychopathy a scientific matter. Such advances have not been forthcoming and though the legislators opted to retain the term in the 1983 Act, the Butler Committee (Home Office, DHSS 1975 para 5.15) had good cause to question the conceptual foundation of the legal definition: 'From both the medical and legal points of view the historical development of the concept of psychopathy has given rise to serious confusion. In essence it was originally a causal theory about hypothetical brain disorder to explain and to bring into one category all those persons who were not 'insane' yet not mentally normal. This theory is no longer held. Later the term was used to describe clinical types but the definitions themselves and the types included in them varied from author to author'.

With such a pedigree, it is perhaps not surprising that, as we shall see, uncertainty and confusion are hallmarks of the admission, treatment, and discharge of the legal psychopath.

None of the definitions we have considered is to be found within the classification system used in the U.K., the International Classification of Diseases (ICD–9) (WHO, 1978). The relationship between the legal definition contained within the 1959 and 1983 Mental Health Acts and the psychiatric diagnosis in the ICD system is complex and needs some explanation. We deal with the definition in the 1959 Act more fully later in this chapter. For the present we need note only that it does not appear in the ICD. The only overlap is that the description of ICD, 301.7, 'sociopathic personality disorder', includes three words that also occur in the legal definition, when it says that people with this personality disorder 'may be abnormally aggressive or irresponsible'. The legal definition thus has no specific meaning in the nosology employed by British psychiatrists. Like the term insanity, psychopathic disorder as defined by law has to be regarded as a legal concept and not a psychiatric one. As Shepherd and Sartorius (1974) point out, 'the relationship between personality disorders, criminality, and antisocial behaviour, raises special problems . . . 301.7 . . . is an unsatisfactory ragbag which is peculiarly susceptible to misunderstanding and misuse'. The confusion has been added to by the adoption of the term psychopathic disorder within the legal framework.

(b) The psychiatric assessment

In the case of mentally ill men liable to be admitted to Broadmoor, the illness itself is the point of reference for the doctor who makes disposal recommendations. It provides for his decision about the man's hospital admission a criterion which is independent of the deed which has brought the man into court.

Mental illness is a term which is undefined in the mental health legislation of the United Kingdom. In recommending detention under the criminal provisions of the Act, most psychiatrists restrict the use of the term to major disorders such as the psychoses. Though there is some dispute between psychiatrists with regard to the nature and cause of these disorders, there is general agreement about what constitutes a psychotic condition. In a mental state examination, the psychiatrist looks for various phenomena, behavioural and verbal, which are associated with disorders of mental function. In the case of a man whose illness manifests itself in an encapsulated delusional system unaccompanied by other symptoms, considerable verbal probing may be required to obtain evidence of abnormality. This type of interview represents the extreme end of a process of examination which can involve, at the other extreme, a diagnosis made largely on the basis of observation of a patient's hyperactive or catatonic behaviour, without any coherent verbal communication taking place between doctor and patient. The man or woman who cannot be defined as mentally ill behaves within normal limits in respect of these functions. However, the Mental Health Acts have, since 1959, allowed people who are neither mentally ill nor handicapped to be detained in a hospital if, in the opinion of the doctors, they are suffering from Psychopathic Disorder.

(c) Legal requirements

The 1959 Act defined the term as 'a persistent disorder or disability of mind (whether or not including subnormality of intelligence) which results in abnormally aggressive or seriously irresponsible conduct on the part of the patient, and requires or is susceptible to medical treatment'. This definition has not been changed fundamentally by the revised provisions of the 1983 Act. It will be seen in Chapter 9 that Broadmoor's consultants have not found the new provisions to make any practical difference.

Thus, under the Mental Health Act, psychopathic disorder is defined as a product of *mind* which leads to or causes aggressive or irresponsible behaviour. When the doctor tells the court that the person under consideration is suffering from psychopathic disorder, he is declaring that the person suffers from a disability of mind that was present before the irresponsible or abnormally aggressive behaviour which is said to result from it.

Before a court can make an offender subject to a hospital order, and before the Home Secretary can transfer a prisoner to hospital, two doctors must present evidence as to his suitability for hospital detention. As a rule, one of the doctors giving evidence to the court will be a prison medical officer and the other, an NHS or special hospital consultant. Their evidence is generally given in the form of written reports on standard forms available from Her Majesty's Stationery Office. If a hospital order with restrictions is to be made, one of the doctors must provide oral testimony to the court.

We carried out a systemic examination of all the forms completed by the referring doctors at the time of the man's trial or before his transfer from prison. They were available for practically all of our 117 men, and were designed to record the reasons and justification for use of the term psychopathic disorder and for the detention of people so described. The law does not require the doctor concerned to make a diagnosis, nor does it state that any particular criteria should be met, but only that two doctors should state they believe the person to be suffering from psychopathic disorder as legally defined, and that the condition 'requires or is susceptible to medical treatment'. In effect, the testimony of the doctors is the proof both of the abnormality of mind and of the need for treatment.

The medical recommendation forms require information to be provided on four points. Only if the court (or the Home Secretary in prison transfer cases) is satisfied on these, can the offender be sent to hospital. The four points are:

1 . Information to establish a persistent disorder or disability of mind.
2. Information to establish that the disorder or disability of mind results in abnormally aggressive or seriously irresponsible conduct on the part of the patient.
3. Information to establish that the disorder or disability of mind requires or is susceptible to medical treatment.
4. Reasons for the conclusion that the disorder or disability of mind is of a nature or degree which warrants the detention of the patient in hospital for medical treatment.

An example of a completed form is reproduced below. It was selected at random, and no changes to it have been made, except where detail would identify the individual concerned.

1. *Information to establish a persistent disorder or disability of mind.*

He is illegitimate, having been brought up in children's homes. Since leaving (these) he has been unable to persist in work and on a number of occasions he has assaulted boys who appear to form his main sexual interest.

2. *Information to establish that the disorder or disability of mind results in abnormally aggressive or seriously irresponsible conduct on the part of the patient.*

On the last occasion (i.e. the offence) he tried to kill the child he assaulted and he does not know why he did this. There appears to be a lack of remorse and failure to appreciate the actual seriousness of his behaviour.

3. *Information to establish that the disorder or disability of mind requires or is susceptible to medical treatment.*

He is an introverted young man who is unable to establish normal relationships with females and who expresses his sexual fantasies in an antisocial and rather dangerous fashion. There is a likelihood that he may respond to therapy in a structured situation.

4. *Reasons for the conclusion that the disorder or disability of mind is of a nature or degree which warrants the detention of a patient in hospital for medical treatment.*

He suffers from a personality disorder and he does not express any remorse about his reported offences. In my opinion he is a potentially dangerous young man and I shall recommend a Restriction Order under Section 65 of the Mental Health Act be ordered by the Court.

The above form contains the average amount of detail provided in such documents, though some are very brief indeed. It also highlights the difficult task with which the doctor is presented. As we have seen the law requires that disability of mind should be present which *results* in abnormally aggressive behaviour. In practice the doctor is presented with a piece of behaviour which is aggressive and *from which* he infers disability of mind. To fulfil the requirements of the law, the doctor in the case presented above has to establish, in the first section of the form, a persistent disorder of mind. The material presented is of course, largely irrelevant and the impression given is that any social abnormality, such as illegitimacy or being brought up in a children's home is 'grist to the mill'. The only detail that is even arguably relevant is that the man had assaulted boys, and that under-age boys formed his main sexual interest. In all four sections, reference is made to this aspect of the man's functioning and, in reality, it forms the only basis for the order being made. The other matters referred to, such as emotional deprivation and employment problems would not, in the absence of the offending, have led to compulsory hospitalization.

In completing forms for mentally ill offenders, doctors are asked to supply information about the symptoms and nature of the illness (see Chapter 2). In the case of the legal psychopath, there is no requirement on the doctors to report on either of these matters. Yet the whole procedure is supposedly a medical one; the questions in the form suppose there to be a medical understanding of what constitutes mind and disability of mind. As far as we could determine from our examination of these forms, doctors recognized what constituted bad or deviant behaviour but gave no account of how or why such behaviour should be regarded as a medical matter. The term 'mind', is essentially philosophical rather than medical; it is none the less employed in the law and on the forms, to 'medicalize' the procedure which turns the badly behaved offender into a legal psychopath and psychiatric patient. The procedure is accomplished without requiring doctors to provide any hard medical information.

Unless there are quite exceptional circumstances, for example the notoriety of a defendant, the evidence of the psychiatrist who states that an offender is suffering from psychopathic disorder is not tested in court. If it were, no barrister worth his fee would have difficulty in demolishing the weak, circular, and retrospective arguments that are produced to establish disorder of mind when no formal illness or mental handicap exists. However, as a rule, lawyers on both sides of the case will have agreed on the defendant's hospitalization before the case is heard and have no interest in putting his legal classification to the test. Neither do judges as a rule seek to examine the doctors' testimony in depth; although the offender is to be sent to hospital by an order of the court, the doctors' testimony (verbal or written) on which that order is based seldom receives systematic or searching scrutiny from the courts.

(d) The need for treatment

All the PD men had been hospitalized because doctors had testified that they required or were susceptible to medical treatment. We therefore examined the court forms for the purpose of seeing what they said about this. Before reporting the results, one preliminary point should be mentioned. The form given to the court is normally accompanied by the doctor's psychiatric report, and it may be thought that the doctors might have dealt with these questions more fully in their reports than in their forms. We therefore checked more than 30 forms against their accompanying reports for the items we were interested in, i.e. those listed in Table 6.2. In only one case did we find that an item had been mentioned in a report and not in a form. Otherwise there were no inconsistencies. There is then no reason to believe that an examination of written reports would have altered our findings.

We examined the forms to see whether they contained information on: (i) susceptibility to medical treatment, (ii) reasons why treatment was required, and (iii) the type of treatment required.

The details are in Table 6.2. In only 23 per cent of the cases did the forms say that the offender was susceptible to treatment, and we recorded what reasons, if any, were given for this belief. In a third of these cases no reason at all was given, but in half, the offender's motivation was referred to. Other reasons (not mutually exclusive) mentioned as evidence for susceptibility were the offender's youth (10 per cent of cases), his failure to respond to imprisonment or borstal (16 per cent of cases), and the previous lack of any trial of psychiatric treatment (4 per cent).

The great majority of men were not described as susceptible to treatment and so, by default, presumably came under the heading of simply 'requiring' it. Occasionally the doctors explicitly pointed to the difference. For example in one case a doctor wrote as follows: 'his behaviour and attitude is so removed from the ordinary as to constitute moral insanity although there is no legal or medical insanity (psychosis). This requires treatment within the meaning of the Act but I do not believe he is susceptible to treatment'. Reasons for believing that the offender 'required' treatment to which he was not described as being susceptible were rarely noted by the doctors on the forms. As a rule it was simply the violent offending that was cited as evidence that treatment was required.

The second part of Table 6.2 shows how often doctors gave any particular reason on their forms for thinking that psychiatric treatment was indicated. Motivation was mentioned in respect of only 22 per cent of the sample. Pre-trial assessment of motivation is of course a difficult process: patients may exaggerate their interest in treatment, and in any case their response to psychotherapy is always uncertain. Yet if offenders who are not mentally ill are to be successfully treated, they themselves must be willing to undertake treatment, since the procedures involved, such as psychotherapy, depend on the patient's co-operation. Our examination of the forms did not suggest that psychiatrists had as a rule treated motivation as being an issue of central importance. Sometimes indeed its absence was cited as a reason for detaining the patient under a court order; for example, in a case where the doctor wrote 'I do not think he will co-operate voluntarily. He therefore requires compulsory detention'. Doctors rarely specified the nature of the treatment they considered to be required; in 76 per cent of cases there was no mention of it. The last section of Table 6.2 shows what treatments were mentioned. In 10 per cent of cases a 'structured milieu' was suggested as being of potential benefit, in 8 per cent of cases psychotherapy was suggested, and in three cases reference was made to the use of anti-libidinal medication.

Our scrutiny of the medical forms for the purpose of seeing what was said about the need for treatment pointed to a fundamental problem which Broadmoor's consultants face in dealing with their PD patients. The great majority of these offenders had been committed to hospital without any indication being given of why they were thought to require treatment, what the treatment should be, or in what way it would affect their propensity to offend. The need of these men for treatment, like their disorder, had simply been inferred from their offending behaviour. The problem is that once such men are admitted, Broadmoor has no choice: they may be untreatable, but if their records are sufficiently serious, they are also undischargeable. The following example illustrates a case of this kind. The man had been in Broadmoor for well over a decade at the time of our research. He had a long history of serious but unexplained attacks on others. The medical form stated: 'He is in need of prolonged medical care, investigation and treatment in order to eliminate the cause of his repeated aggression . . . Until this has been achieved, . . the public stand in need of equally prolonged protection from any further consequences of this aggression'. There was no suggestion in the report about *how* Broadmoor was to 'eliminate' the causes of aggression. When we interviewed this patient's consultant, he said that the case had remained baffling. The offences still remained unexplained, psychotherapy had not proved helpful, and no one knew what to do to eliminate the problem.

Finally, mention should be made of a group of men who had explicitly been sent to Broadmoor to prevent their release into society rather than for curative treatment. In these cases, there had been no pretence that treatment could be expected to change the patient, but it was thought Broadmoor was the best way of providing long-term humane confinement or 'care . . . under medical supervision' to use the legal term of the Mental Health Act. The following is an extract from the admission report of a repetitive sex offender who at the time of our research had spent over 12 years in Broadmoor. The special hospital consultant recommending admission wrote as follows: 'He undoubtedly suffers from a disability of mind within the meaning of the Act. I have grave doubts as to whether his disability requires or is susceptible to treatment. Past treatment has been ineffectual and experience leads one to expect very little benefit in this type of case . . . (but) the matter of danger to the public must be considered . . . The reasons for sending this man to hospital would be more for his confinement than for his treatment.'

It is not likely that the DHSS would nowadays admit a patient on the basis of such a report (see Chapter 1) but since patients like this have been admitted over the years, they are part of today's population.

(e) Persistence

The legal category of psychopathic disorder requires, as we have seen, a persistent disorder or disability of mind to be established. Yet in some cases the cateogry was used for men who had no previous history of offending and who had, before the admission offence, displayed little evidence of any disorder of mind, let alone a persistent one. Sometimes such men had broken down under stress and behaved in a totally unexpected and unforeseeable way. Psychiatrists who carry out pre-trial examinations may interpret such behaviour in a variety of ways. Some may see it as a transient reaction to stress, not warranting psychiatric treatment, whereas others may choose to regard it as coming within the Mental Health Act category of psychopathic disorder (Dell, 1984).

In one case of this kind the trial medical reports revealed a variety of views about a family homicide. There was no disagreement about the offender's unremarkable and non-violent history, only about its interpretation. One consultant thought that the violence was to be seen in terms of a depressed reaction to stress, but he did not think that there was present any mental disorder that required psychiatric treatment or came within the Mental Health Act. Another consultant took a similar view, attributing the offence to a 'transient condition' that had arisen in consequence of 'severe emotional stress': he too found no evidence of mental disorder. Nevertheless other consultants were prepared to say that the man's unhappy family relationships were evidence of an abnormal personality that amounted to psychopathic disorder within the Mental Health Act, and he was made subject to a hospital order. Once again, it was the offence and not any independently existing mental disorder that led to the man's detention: unhappy family relationships are not of themselves considered by psychiatrists to warrant compulsory hospitalization. In Chapter 8 we will be looking at some of the discharge issues that occur when the legal category of psychopathic disorder is used for men who committed isolated offences in transient circumstances.

MAJOR CHANGES IN DIAGNOSTIC CATEGORIES

Of the 117 men admitted under the category of psychopathic disorder, twenty (17 per cent) were diagnosed by their Broadmoor doctors as psychotic when our study was undertaken. Only three of the twenty had been reclassified as mentally ill under the Mental Health Act.

A comparison was carried out between those men who changed status and those who did not. It showed that over a third of the former (compared with 12 per cent of the latter) had at some point before the Broadmoor admission been described as psychotic, and that 60 per cent had previously been psychiatric in-patients. In many cases then, the PD men who were later diagnosed as psychotic did not first develop their illnesses in Broadmoor but had already given evidence of them before they came.

PREVIOUS HOSPITALIZATION

Of the 117 men, 48 (41 per cent) had previously been in-patients in a psychiatric hospital, usually on a voluntary basis. Only six of these men had ever been detained under civil procedures and all six had been held under the mental illness classification. It is worth pointing out that although under the 1959 Act, persons under the age of 21 could be detained under civil procedures for psychopathic disorder, not one of the men in the sample had ever been detained in this way.

BREAKDOWN OF THE PSYCHOPATHIC DISORDER GROUP BY AGE AT ADMISSION

In considering men admitted under the legal category of psycho-pathic disorder we face the problem of providing an adequate account of the heterogeneous nature of the group without reverting to indi-vidual descriptions. Diagnostic groupings were impossible to make since 85 per cent of the men were simply described in their trial medical reports as suffering from 'psychopathic disorder'. For the purpose of describing the population in some more detail, we therefore divided it into sub-groups by age at admission. We had supposed ini-tially that the older men in the population would be those who had been in Broadmoor longer, but a correlation analysis showed that this was not so and that length of stay in Broadmoor was not related to age. The older men were not the young men grown old.

We divided the PD population into three groups on the basis of their age at time of admission to Broadmoor. The first group, 35 per cent of the population, comprised those men who had been admitted between the ages of 16 and 20, the second, 42 per cent of the total, was made up of men admitted between the ages of 21 and 29, and the third, 33 per cent of the population, consisted of men who were thirty or older on admission. The precise cut-off points were chosen to create groups of approximately similar size.

When we compared the groups, they were found to differ as regards their criminal histories. Whereas juvenile offending was relatively rare among the older entry men, it was the norm for men who were under 21 at the time of admission and the intermediate age group was intermediate in regard to juvenile offending. The older entry men had much more experience of prison, 74 per cent of them having served at least one prison sentence as compared with 24 per cent of the young entry men. The older men had more frequently been admitted after sex offences. They also differed from the other groups in that they had more previous convictions for arson and indecent assault, but not for larceny.

Another difference between the groups related to their results on the MMPI. Comparison of the age groups on the scales derived from this test revealed one major difference in that the youngest group produced a significantly higher mean score on the Hypomania scale (McKinley & Hathaway, 1956). The two older age groups produced mean scores on this scale which were within normal limits whereas the young men's group score was well into the pathological range. The scale contains 48 items, some of which relate to expansive mood and social extraversion. Carson (1969) described people scoring highly on this scale as: 'generally outgoing and uninhibited. They tend to become easily offended and may be seen as tense and hyper-active'. Moreover, as the scale score increases, 'there is increasing likelihood of maladaptive hyperactivity, irritability and insufficient inhibitory capacity.' The term impulsive might best apply to the general characteristic being described. Two other individual scale scores also distinguished the groups, the first of these being the PD scale (McKinley & Hathaway, 1956) in which the older age group scored significantly less pathologically than the other two age groups. Secondly, on the hostility scale (Foulds & Cane, 1967) there was a trend for the expression of hostility to decline over the three age ranges.

CONSULTANTS' ADMISSION MEETINGS

As explained earlier, Broadmoor consultants usually go to see pro-spective patients before advising the DHSS on whether to make a bed available: the cases seen are then discussed at a weekly admission meeting of medical staff. By attending such meetings, we learned of the very different views that the various doctors held about the advisability of accepting offenders under the legal category of psycho-pathic disorder. We also learned how the backcloth to the discussions was the presence in Broadmoor of men who had been admitted under the PD category in times when less attention was paid to 'treatability' than it is today. The cases of patients who were untreatable, and because of their serious records, undischargeable, were frequently cited as cautionary examples.

The treatability of potential patients was the issue on which the discussions invariably focused. In one case the prospective patient was in his forties and had a long history of aggressive sexual offences. The doctor who had been to see him favoured admission, but others were strongly opposed. The first question discussed was: 'What treatment could Broadmoor offer?' Those who favoured admission said 'Bring him in, . . . and see where we get to'. This, as we have seen, was in effect how most of Broadmoor's psychopathic disorder patients had been admitted: the reports on them had rarely given any indication of why or what treatment was expected to prove helpful. But the 'bring him in and see where we get to' approach was wholly unacceptable to other Broadmoor doctors. Before they were prepared to consider admission, these consultants wanted expert advice on whether psychotherapy could be expected to help the man. They pointed out that such advice was not available, since the psychotherapists, though they often attended the admission meetings, never went to assess potential patients themselves.

The doctors who opposed admission in this case also emphasized how the gravity of the offender's record would make him difficult to release. They argued that even if the patient did well in therapy over two or three years, his record would be an impediment to discharge: he was likely then to remain in Broadmoor long after useful treatment had come to an end. 'If we take him we are lumbered with him.'

In cases like these, some consultants argued forcefully that it was better for offenders to be given prison sentences rather than hospital orders and for them then to have the opportunity to come to Broadmoor for treatment on transfer. It was argued that this ensured that men who proved to be untreatable or who had come to the end of useful treatment could be sent back to prison, where the length of their detention would be determined by the judicial tariff (Mawson, 1983). Other consultants pointed out that on discharge, no after-care or recall was possible if the transferred prisoner had been given a determinate sentence by the court. They also argued that it was the indeterminate nature of the hospital order that often forced Broadmoor patients to recognize that their only hope of getting out was to change.

At another admission meeting we attended, the possible acceptance of a much younger man, also a sex offender, was discussed. All the doctors agreed that he was potentially dangerous, that he did not need maximum security, that no other hospital would take him and that if he did not go to a special hospital, he would get a prison or borstal sentence of about two years.

Discussion centred upon what treatment might be available in Broadmoor and how the youth would respond to it. Some of the doctors expressed deep scepticism as to whether the hospital had any appropriate treatment to offer to an offender who suffered from no demonstrable mental disorder other than his propensity to offend. 'What is this wonderful treatment? Look at X (one of Broadmoor's long-stay sexual offenders), what have we ever done for him?' All recognized that there was great uncertainty about the patient's response to whatever treatment might be offered him. The consultant who had seen the patient nevertheless favoured admission: 'We don't know whether treatment will work or not . . . . but he requires it and that's enough . . . he's 17 and he hasn't had in-patient treatment before . . . even if we have little to offer, if we don't do this for him no one else will'. Strongly influencing this consultant was the knowledge that if the youth did not come to Broadmoor he would get a penal sentence. 'If we write him off, he will be criminalized before he is 18.'

Other consultants strongly opposed admission. There was no evidence, they pointed out, that the patient was susceptible to therapy: 'How can anyone require treatment to which he is not susceptible?' These doctors argued that the offender's stay in Broadmoor would inevitably be much longer than any prison sentence he faced. They also referred to the damaging stigma of the Broadmoor label that he would receive, more damaging than any prison record. They therefore argued that he should only be exposed to these adverse factors if there was a real likelihood of benefit from treatment. Other consultants dismissed such issues of criminal justice and civil liberties as irrelevant: 'It shouldn't be a consideration.'

An issue that arose periodically in the meetings was the value of after-care provisions attached to restriction orders. Unlike the restricted patient, an offender with a determinate sentence who is transferred to Broadmoor does not, on release, become subject to compulsory supervision or liability to recall. Sometimes it was this difference that swayed the consultants in the admission recommendations they made. In one case in our sample, a young man wounded a woman in a sexual assault. A Broadmoor consultant went to see him and reported that he was of low intelligence, that it was his first offence and that a borstal sentence with the possibility of treatment would be the best disposal. However the doctor later reconsidered the matter and recommended admission to Broadmoor on the grounds that the youth could then get long-term supervision under a restriction order. Such an order was duly made. The effect was that in the interests of post-release supervision, the man was detained for much longer than would otherwise have been the case. Had a Borstal sentence been given, he would have been out of custody within one to two years: as it was, he was still in Broadmoor eight years after admission.

THE PSYCHOPATHIC DISORDER GROUP IN CONTEXT

An exercise similar to that carried out for the mentally ill group was conducted to establish how the men in the Broadmoor sample differed from those admitted under the same legal category to local hospitals. As before, use was made of the information which had been collected by Robertson and Gibbens (1979) in their fifteen-year follow up of people given local hospital orders by courts in 1963/64. The comparison is presented in Table 6.3.

The figures on the left of the Table refer to PD men sent to local hospitals under hospital orders without restriction. The middle column gives details of PD orders to local hospitals made under restriction and the third column represents the restricted order group in the present study.

The most interesting difference between the group rests in their respective records of previous in-patient care. In both the local hospital populations, only a small minority of men (14 per cent and 18 per cent) had not received in-patient care before being given their hospital order. In the case of the Broadmoor group the majority (58 per cent) had never previously been in-patients. Before they committed the index offence, hospitalization had been considered appropriate for the vast majority of local hospital PD admissions, but this had not been the case for the majority of the special hospital intake.

This finding has its correlate when we consider the gravity of the admission offence. Overwhelming every other consideration, this is what distinguished the special and local hospital populations. The great majority of the special hospital patients, 71 per cent, had committed crimes of violence, including 30 per cent who had commited homicide. In contrast, of restricted patients in local hospitals only 6 per cent had been involved in aggressive offences, none more serious than assaults, and 50 per cent had been hospitalized following a larceny offence. So overwhelming a factor is the gravity of the admission offence, that a variable which might be judged to reflect suitability for treatment, namely a previous history of psychiatric in-patient care, is inverted in the special hospital group. The fact that Broadmoor admits so large a percentage of PD men who have never previously been psychiatric in-patients demonstrates that its criterion for judging the suitability of such admissions is fundamentally different from that which operates within conventional psychiatric hospitals. What happens to these men once admitted is the subject of the next chapter.

7 Treatment of non-psychotic men

INTRODUCTION

This chapter discusses the treatment of the patients in our sample who were not described by their Broadmoor doctors as suffering from psychotic illness. There were 106 such men. The group consisted of (1) 97 of the 117 men who had been admitted as psychopathically disordered (20 of the 117 were, as noted in the previous Chapter, described as psychotic), and (2) nine non-psychotic men (described in Chapter 2 and Appendix 1) who had been admitted under the legal category of mental illness. All but five of the 106 men were interviewed: two declined to participate, and three were discharged before we could arrange to see them.

Details of the diagnoses given to us by the consultants are shown in Table 7.1. It will be seen that 70 per cent of this population were given a diagnosis of psychopathic disorder or unspecified personality disorder. Another 21 per cent were described as suffering from specific personality disorders, and the remaining 10 men were diagnosed variously as neurotic (4), sexually deviant (4), epileptic (1) or addicted to drugs (1). It would have been possible to refer to the group as 'personality disordered, neurotic, etc.', but we have chosen the equally accurate but less cumbersome title 'non-psychotic'. Sometimes we also refer to them as not mentally ill, and obviously that does not imply that we think there is nothing wrong with these men.

It would have been desirable, for the purpose of looking at the treatment they received, to have described the non-psychotic group diagnostically, but this was not practicable as so many of the diagnoses were no more detailed than 'psychopathic disorder' or 'unspecified personality disorder'. We therefore decided for the purpose of looking at treatment, to group the men in terms of age. As seen in the preceding chapter, there were distinct differences between the men who came to Broadmoor in their youth, and those who came later in life. The 106 men were therefore divided into three groups on the basis of their age at admission: patients aged 21 or less (43 men), those aged 22–29 (32 men) and those aged 30 or more (31 men). The men in the three groups had been in Broadmoor for similar lengths of time, on average 8.1 years (s.d. 4.5).

The groups were compared on all the variables we had collected, in order to see how far the differences in the treatment they received related to considerations other than age. Attention should be drawn to only one major area of difference, which can be seen in Table 7.2: the older men had more often been admitted after sexual and arson offences, and they had more previous convictions of this kind. About half of each age group had come to Broadmoor after a sexual assault, but the oldest group contained more men with persistent records of this type of behaviour. As far as treatment was concerned, there was an important difference in how the sexual problems of the groups were seen by the consultants. The behaviour of the younger men was generally regarded as arising from problems of immaturity or conflict, whereas the offences of the older men were seen as demonstrating established sexual deviations.

TREATMENT IN BROADMOOR

(a) **Medication**

The great majority of the group were not treated by medication. At the time of our research only 15 of the 106 men were taking psychotropic drugs: they came mainly from the older groups. Of the 15 men, six were on major tranquillizers, six on minor tranquillizers, and the others were on antidepressants. Most of those taking medication said that they were content to be doing so, and that they thought it had helped them.

Of the great majority of men who received no medication, a number commented with regret that their disorder was not amenable to drugs: for example, a paedophiliac man told us: 'There are no medicines that can cure my problem'. Sex offenders in particular were liable to raise the question of medication or 'chemical castration'. In a few of these cases, the issue had in fact arisen at the time of their trial. One of these men had a history of offences against children and his trial reports had stated that his urge to offend in this way might be remedied by specific treatment: 'He wants to be cured of his abnormal impulses. This to some extent may be possible with special hormones'. When the man came to Broadmoor, however, he found that no such treatment was forthcoming, nor did the hospital have any other help to offer. He continued to be troubled by sexual thoughts about children and eventually asked his lawyers to request that he should be given libido-suppressant drugs. The Broadmoor authorities replied that this was not possible because of the ethical and legal problems of giving such preparations to detained patients. The patient could see no way out of his predicament 'I wish I could have a brain operation . . . or castration . . . I've asked for it, but they said we don't do it . . . There is no treatment for me'.

(b) **Individual psychotherapy**

At the time of the research, 15 men (nine of them in the younger age group) were having individual psychotherapy. All but four were being seen by visiting therapists: the others by Broadmoor psychologists or by the speech therapist, who was also a trained psychotherapist. It can be seen in Table 7.3 that 43 per cent of the men had received individual psychotherapy at some point during their Broadmoor careers, the younger patients having had it considerably more often than their elders. Treatment normally consisted of a weekly session of one hour's duration. The average number of months the men had spent in individual psychotherapy was 21.6, but as the standard deviation of 19.3 indicates, this variable was widely distributed.

When we asked patients for their opinions about the value of individual psychotherapy, about a quarter said they had not really found the treatment useful, another quarter said it had been quite helpful, and half rated it as very helpful. The value of learning to talk and communicate was often referred to, but the most common comment was about how psychotherapy had helped the participants to understand themselves and their behaviour. One patient said 'It was very very good; I gained a lot . . . I was able to talk about anything . . . I can (now) understand the problems that got me into trouble'. Some men referred to the quality of the relationship with their therapists, and felt this had been the key to progress: 'It did one thing for me. I didn't trust anyone before . . . I thought everyone was out to get me . . . I came to understand that not everyone was'.

(c) **Group therapy**

At the time of the research, 14 men (13 per cent) were having group therapy. Again, most of them (11) came from the youngest group. Overall, 71 per cent of the men had taken part in group therapy at some time during their stay and, as Table 7.3 shows, it was the youngest group that had most often been involved.

Groups were taken by psychologists, by the visiting psychotherapists, or by the speech therapist. As a rule, a member of the nursing staff also took part in each group. Sessions were held weekly and lasted an hour. The average number of months that men had attended groups was 23.1 (s.d. 19.7), but that figure did not necessarily represent time spent on the same group, for it was not unusual for patients to have attended two or more different groups during their stay. Almost a quarter (23 per cent) of the men who had participated in group therapy had attended for no more than six months.

Group therapy was seen as being considerably less valuable than individual psychotherapy: only 28 per cent of the participants had found it very helpful, and almost half (49 per cent) said it had not really been of any use to them. The main reason for this, given by men of all ages, was that they felt concerned about confidentiality. They said that they had not been prepared to talk freely in their groups because they knew (and often they cited examples) that damaging and embarrassing material was liable to be repeated outside. Groups thus became centres of conversation rather than of therapy. But as such, they were still seen as having some value: a number of men who said that groups had not helped them with their basic problems nevertheless thought that they had helped to ease tensions and conflicts of ward life.

Among the minority of men who were enthusiastic about group therapy, appreciation was warmly expressed. 'You stop feeling a freak . . . you find others are like you'. 'Terrific . . . the therapist didn't put words into your mouth . . . he went back to the beginning . . . helped you to understand why'. 'In the group you can discuss whatever you want . . . you're not interrupted by brothers and sisters . . . it's helped me to understand where I went wrong'. 'You become aware of things . . . scales fall off your eyes, wax out of your ears'.

(d) Social skills groups

Twenty-three men (22 per cent) had taken part in a social skills pro-gramme at some time during their stay: again, the majority (70 per cent) came from the youngest age group. At the time of the research, three men, all in the youngest group, were taking part in such a course.

The programme was usually conducted by psychologists and con-sisted of weekly group sessions of about an hour's duration, carried on for a period of a few months. The object was to help patients to improve their social skills: for example, their ability to talk to strangers and to initiate appropriate contacts with the opposite sex. Role playing and videotaping were used. The work has been described by Crawford (Crawford & Allen, 1979; Crawford, 1978).

Opinion about the value of these groups was generally favourable. Three-quarters of the men rated them as helpful, including over half who described them as very helpful. Participants felt they had been given insight into how others saw them, and that they had acquired useful information about appropriate ways of behaving.

(e) Other treatments

Behaviour modification programmes run by psychologists had been attended by a small number of patients. Six men had at some time been involved in relaxation therapy, four men in sex behaviour modification programmes, six in anger control sessions, and four in other programmes. The majority of these men reported that they had found these sessions useful. Only two men were receiving this type of treatment at the time of our interviews.

Sex education courses were periodically run for the younger men. They usually lasted for a few months and were held on a group basis, at weekly intervals. Altogether 23 of the men (22 per cent) had attended such a course during their time in Broadmoor: all but one came from the youngest age group. Two-thirds of these men said they had found the courses helpful, a third having rated them as very helpful.

SUMMARY OF TREATMENTS RECEIVED

Table 7.3 summarizes the treatment data on Broadmoor's non-psychotic population. It shows what specific treatments the men were having at the time of the research, a snapshot picture, and it shows what specific treatment each patient had received in the course of his stay. Three features are worth noting. First, at the time of the study, only a small minority of the men were taking part in any specific treatments: 14 per cent were receiving individual psychotherapy, 13 per cent were having group therapy, and 8 per cent were taking part in one of the other non-physical treatments. Two-thirds of the non-psychotic population were not taking part in any specific psycho-therapeutic programmes.

Secondly, the treatment programmes were largely concentrated on the youngest age group. Of the 35 men who were taking part in specific treatments at the time of our research, 25 (71 per cent) came from the youngest group.

The third point of interest is in the overall treatment history of the population. First, group therapy was the only treatment procedure of which a majority of men from each age group had experience, and half of those who had had it did not find it helpful. In the other forms of therapy, the oldest group had only rarely participated, and 36 per cent of them (as compared with 7 per cent and 9 per cent of the other two groups) had taken part in no specific treatments at all. Secondly, it was apparent that in all age groups, for most of their time in Broadmoor, the men were engaged in no specific treatments.

We calculated a figure for the total time spent in all non-physical treatments. As would be expected, this correlated with the length of time the men had been in Broadmoor ($r = 0.26$ $p < 0.004$), but it was interesting that the correlation with age at admission was more statistically significant ($r = -0.37$ $p < 0.0001$). Because of the relationship with length of stay, we calculated 'treatment time' as a proportion of time in Broadmoor (Table 7.3). Our calculation assumed that treatments were consecutive rather than concurrent, which was, of course, often not the case; men might be attending social skills training at the same time as they were attending group therapy. Furthermore, our calculation treats 12 months in therapy as 12 months of treatment and not as the actual time spent in therapeutic sessions, which would at most have amounted to a few days real time. Our figure for time spent in treatment is therefore greatly exaggerated. Nevertheless, it serves both to illustrate the difference between the age groups and to show how little of the men's time was spent in specific treatment. In no group did the proportion of time in treatment reach 50 per cent; for the youngest men, it was 46 per cent, for the middle group, 32 per cent and for the oldest, 16 per cent. On average, these non-psychotic patients had spent 8 years in the hospital, and during at least two thirds of this time, the only treatment they received was that of 'being there'.

MILIEU THERAPY

Hospital patients are exposed to more than the specific treatment programmes in which they take part. Being an in-patient means being part of a residential community, interacting with nurses, doctors, and fellow-patients, and living the kind of life that the institutional organization determines.

In the 1959 Mental Health Act, the phrase 'medical treatment' is defined as including nursing, care and training under medical supervision. (The 1983 definition is similar.) We have already noted in Chapter 3 that doctors regard life in the hospital community as providing a form of treatment in itself, milieu therapy. They consider the experience of patients living together in a controlled and secure environment to be therapeutic, in that it enables them to mature and to learn what forms of behaviour are acceptable. Thus 'being in Broadmoor' is regarded as treatment, and for some of the non-psychotic men it is, as we have seen, the only treatment offered.

A few of the men said to us that it was life in Broadmoor rather than any specific treatment that had helped them. One man echoed the psychiatric theory when he told us: 'Everything you meet inside, you'll meet outside . . . It's the environment where you learn'; and another man reported: 'My only treatment was environmental . . . Just being here has given me time to mature'. However, the idea that being in Broadmoor was in itself treatment was not generally accepted by the patients. More common was a response like the following: 'Milieu therapy? What it means is that if you lock a man up he can't get into trouble'.

PATIENTS' VIEWS ABOUT TREATMENT IN BROADMOOR

(a) Satisfaction with treatment

We asked the men if they were satisfied with their treatment. Almost half (47 per cent) said that they were not, a third said that they were, and 14 per cent were ambivalent. The replies did not vary across the age groups, but were strongly related to length of stay: long-term patients were the least satisfied with their treatment.

When the men were compared with those who had been diagnosed as psychotic, we found that significantly more of the latter expressed satisfaction with their treatment, 54 per cent, as compared with 34 per cent ($p < 0.02$). More of the psychotic men felt that something relevant was being done for them. Many of the non-psychotic men felt that years in Broadmoor had left their problems untouched.

This emerged clearly when we asked the men to explain why they were dissatisfied with their treatment. The great majority (75 per cent) of the dissatisfied non-psychotic men gave the same reason: they felt that their problems had not been tackled, and that no suitable treatment programmes had been devised for them. They could not see 'being in Broadmoor' as an adequate treatment, and were unhappy at the lack of specific therapies directed to their needs. One man told us (accurately enough, as his records showed): 'Nobody knows what to do with me . . . Until they decide I've mellowed, I'll just sit here'. Some men referred to the mysterious disorder which they were supposed to be suffering from. 'Personality disorder is a frightening mesh to be caught in . . . The doctors say 'Get on and get well'. But there are no criteria on how to do it'. Another man put it more bluntly: 'Treatment for psychopaths? It's like the emperor's clothes'.

(b) Was it fair?

The non-psychotic men knew that for them the alternative to Broadmoor would have been imprisonment, and as we shall see, they often thought that Broadmoor was a better place in which to be serving their time. When we asked whether they thought what had happened to them was fair, the majority (59 per cent) said they thought it was; 17 per cent said that they thought it unfair, and a fifth thought they had been held too long for it to be fair. The answers to our question about fairness showed that men whose offences had not been serious enough to qualify for indeterminacy (life imprisonment) in the penal system felt to an overwhelming extent that they had been unfairly treated by being sent to Broadmoor: 86 per cent said that what had happened to them was not fair. Only 25 per cent of the men whose offences could have led to life sentences gave this reply ($\chi^2 = 26.08$ 2 df $p < 0.001$).

A number of patients said they thought the time they had served was fair in retributive terms, but not in therapeutic ones. 'Yes, fair on the grounds of what I've done. But not on the grounds of what I'm liable to do' is how one man put it. Here the man touched on a problem that the doctors were reluctant to admit. The theory of sending mentally disordered offenders into hospital under the Mental Health Act is that they should be moved out as soon as they get better. But in practice, for offenders who are not mentally ill, an element of 'time' has crept into the procedures (cf. Grounds, 1987a) and patients are of course well aware of this. A homicide offender gave us his assessment very clearly. He had been in Broadmoor for four years, and said that considerably more time would have to pass before his release was seriously considered. 'You have to serve the right length of time. . . . What is not fair is that they *say* you're here until you get better, and that time doesn't count. . . . Time does count, although they say it doesn't. With the psychiatric jargon, they can plant anything on you, and so they don't ever have to say you haven't done enough time. . . . They can always find a psychiatric reason'.

(c) **Prison or Broadmoor**

We asked the men whether in the light of their experience, they would rather have served a prison sentence or come to Broadmoor. All had had experience of prison while on remand and most had served prison sentences in the past. A minority, 37 per cent, said they would choose prison. Twelve per cent were ambivalent, saying they would choose Broadmoor for its conditions and treatment, but prison for length of sentence. Half the men said unequivocally that they would choose Broadmoor. Men in the oldest group were more likely than others to say they would rather have gone to prison, no doubt because Broadmoor had so much less to offer them. Over half of the oldest group, 52 per cent, gave this response, as compared with 27 per cent of the youngest and 38 per cent of the intermediate age group ($\chi^2 = 4.20$ 2 df $p < 0.12$).

Two main reasons for preferring Broadmoor to prison were given. One was that Broadmoor was more comfortable, less restrictive, and the pay was far better: 'If you're doing a long time, you might as well be comfortable'. The second reason, most often mentioned by the younger men, related to the treatment possibilities Broadmoor offered: 'In the nick you're left to your own devices, just waiting for the release date. Here you put your time to use'.

The men who said they would rather have gone to prison often believed, not necessarily accurately, that they would have served shorter sentences there, or at least would have had more certainty about the length of their detention. Most of the non-psychotic men (73 per cent) had come to Broadmoor after offences for which they could have got life sentences, but there was a firmly held view that in prison 'life' meant nine years or so, and that those serving it therefore knew where they stood.

A variety of other reasons for preferring prison to Broadmoor was given. A number of men referred to the stigma of Broadmoor, more permanent and more dreadful than the stigma of prison. Others referred to their fellow residents: 'In prison you're mixing only with sane people . . . you're not among the ill'. The therapeutic framework, which some people gave as the reason for preferring Broadmoor, was given by others as a reason for preferring prison. 'The nick is an unpleasant place . . . Broadmoor is easier . . . but prison is more honest, there's no pretence about rehabilitation'. Another man put it more strongly: 'In prison they say you can do time the easy or the hard way. Here, they want you to change completely . . . you have to say "yes doctor, you're right" . . . you have to bow down to get through and out'.

Some patients were ambivalent about whether they would choose prison or Broadmoor. One man said 'I'm not sure . . . it hasn't all been bad here. They've tried to help . . . It would give me more pleasure to say that I preferred prison. No one wants to say that they need psychiatric treatment . . . I'd prefer to think of myself as a macho prisoner and not as a patient'.

(d) **Had Broadmoor helped?**

We asked the men whether they thought being in Broadmoor had helped them in any way. Three-quarters of them said it had. The age groups differed significantly in the way they replied: almost 90 per cent of the two younger groups, as compared with 52 per cent of the oldest, said that they had been helped $(\chi^2 = 15.23$ 2 df $p < 0.001)$.

Patients who responded positively to the question were asked in what way Broadmoor had helped them. Almost half (45 per cent) said that they had learned more about themselves, and had gained in self-understanding. Thirty-six per cent said that it had helped them to get on better with other people, and a third said it had taught them better self-control. Many other matters were mentioned, not all of them directly related to Broadmoor's role as a hospital. For example, some patients said they had been helped by learning that others had difficulties similar to or even worse then their own. A man with a drinking problem said simply: 'I feel much better without alcohol. It's enabled me to sober up and see life in reality'. Several men said that Broadmoor had enabled them to accept their homosexuality. Others referred to the opportunities the hospital had offered them in the field of education, and the difference this had made. One man who had gained considerable qualifications during his stay told us 'I know my strengths better now . . . I didn't think I had any strengths before . . . I felt the lowest of the low'.

There was no doubt that some men had found Broadmoor profoundly helpful: 'It's given me a chance to start life again'; 'I wish years ago I could have come to a place like this. Then perhaps all this wouldn't have happened'; 'It's done me a lot of good . . . I wish I could have had it without doing what I did'.

(e) **The worst things and the best**

We asked each man what he thought was worst about being in Broadmoor. The replies did not vary between the different age groups, and most frequently they related to the indeterminacy of the detention: 'Not knowing what you can do to prove to the doctors that you are ready to go out'; 'Not knowing is the killer'; 'Never knowing where you stand in relation to time and everything else'. This last comment touched on a related problem to which some men referred: the lack of factual information about aspects of their detention. For example, patients sometimes had case conferences, or they were told that attempts were to be made to transfer them, and then heard no more. They found it hard to be left speculating: 'Is my doctor slogging his heart out over it, or has he forgotten all about it?'.

After the indeterminacy, the lack of freedom and the excess of control were most often mentioned as the worst thing. Other issues raised included the lack of treatment: several sexual offenders said that for them the worst thing was to be detained and to be seeing nobody about the problems that had led to their detention. One man referred to something which many had mentioned in other contexts; that 'everything here is in slow motion': progress towards moving people out seemed to occur at a snail's pace. A number of men referred to the social problems of institutional life: 'It's lonely in a crowd. All we have in common with each other is that we're here'. Lack of privacy was a common complaint, as was concern about becoming institutionalized: 'People do everything for you, you have no responsibilities'. Some patients, mostly from the youngest group, thought that the worst thing about being in Broadmoor was the distance from their homes and ensuing problems about visits.

One problem mentioned by a number of men has already been referred to: the fear and anxiety they felt about living together with psychotic people. One man referred to psychotic patients on his ward: 'They hear voices, and you can never tell when they'll turn round and wrap a chair round you for no reason'. Another man told us: 'It's very frightening. I'm scared of becoming like that if I'm in contact with them long enough'.

We asked each patient what was best about Broadmoor. It was not possible to categorize the answers, as too many men answered negatively, but there were some interesting positive responses. The importance of sports and outdoor facilities was reflected by the number of younger men who said that the best thing was 'football on Wednesdays and Saturdays. . . . We can get out into the fresh air'. A number of men said the best thing was being left alone to work their way through the system: 'Being left reasonably alone to my own devices, to work my way through the system, jobwise, parolewise, roomwise'. Men who felt they had benefited from the educational opportunities or from treatment sometimes mentioned these. A sex offender said 'You're treated as a human being no matter what offence you've committed'.

The variety of responses to our questions about what was best and worst demonstrated the heterogeneous nature of the population. It also reflected the fact that for some patients being in Broadmoor was a purely custodial experience with no relevant treatment and little reason for hope, whereas for others it was a therapeutic one, at the end of which the patient envisaged returning to the world better fitted to cope with life. So an elderly patient who had been in for 13 years and had little to look forward to in the way of treatment, said to us 'There's no good points about this place. . . . It's a nightmare. . . . Locked up day after day with the same mad people'. Whereas a much younger man in another part of the hospital said: 'It's like being taken from one family to another. I didn't think it could be so good'.

(f) Ready to go?

We asked each man whether he felt ready to go out: 18 men (18 per cent) categorically said they did not feel ready. When we asked for the reasons, the younger men most often said that they were in the process of receiving treatment (usually psychotherapy) and wanted to complete it. One such man told us there were still 'things to be cleared up and learned before I feel fit to go out. . . . I'm only half way there'. In the two older groups, fears about the future were usually given as the reason for not wanting to leave; four men said they were afraid they might reoffend, and another four expressed fears about their ability to cope with the outside world. Two men said frankly that they had settled and did not want to leave. One of these had been in Broadmoor for fifteen years and told us: 'I've become institutionalized . . . things here suit me. It's comfortable . . . easy going . . . no hassle. . . . There's lots of tramps out there who would queue up to come here if they knew about it'.

There is no doubt that Broadmoor provides a wanted and welcome asylum for a small group of non-psychotic men. It includes some of the compulsive sexual offenders who fear that they would be unable to control themselves if released. It includes men who have never been able to handle the problems of life in the community and others who, because of their explosively violent personalities, have found it impossible to lead acceptable lives outside. One such man, for whom Broadmoor had at last provided an environment where he could live and work, expressed his appreciation: 'They leave you alone to get on with things . . . no shrieking women and no nagging children'.

BEHAVIOUR IN BROADMOOR

Seven men (7 per cent) were recorded as having been involved in a violent incident against staff members, and 25 (24 per cent) had been involved in some kind of aggressive disturbance with fellow-patients. In terms of hospital management the non-psychotic men were not an unruly group: as was seen in Chapter 3, mentally ill patients caused considerably more disturbance, and more psychotic than non-psychotic men had spent time in the intensive care unit (35 per cent compared with 14 per cent, $p < 0.001$). However, the non-psychotic men were more often involved in planned breaches of hospital discipline: 13 of them had been in trouble for behaviour like attempts to escape or trying to hide offensive weapons.

DOCTOR AND PATIENT RELATIONSHIPS

We asked the men how well they thought their consultants knew them, and we asked the consultants how well they thought they knew each of their patients. The consultants replied that they thought they knew their patients well in 81 per cent of cases, not well in 18 per cent, and hardly at all in 1 per cent. The patients thought that their consultants knew them well in 42 per cent of cases, not well in 38 per cent and hardly at all in 17 per cent. The majority of men were thus unhappy about one important element in the doctor-patient relationship, and many comments were made about the need for better personal contact: 'They should spend more time on the wards, more time talking to patients. . . . They should be more involved'; 'The psychiatrist should be the strongest not the weakest link in the treatment chain'; 'Contact with doctors is minimal'; 'The main problem here is that you don't get to see the doctor.'

CONSULTANTS' VIEWS ON THE RELATIONSHIP BETWEEN TREATMENT AND OFFENDING

We asked the doctors what they thought had caused each patient to offend, and what could be done about it. For the psychotic men, as we saw in Chapter 3, the consultants usually said that their illnesses had caused them to offend, generally because delusions had driven them to commit acts of violence. When we asked the doctors the same question about their non-psychotic patients, the responses were very different; as a rule they did not ascribe the offences of these men to specific disorders. The following three examples give some flavour of the doctors' responses to our question 'What do you think causes him to offend and what do you think can be done about it?'

In the first case, the patient was an arson offender, who had been in Broadmoor for six years. The cause of his firesetting was said by his consultant to be due to his inadequate personality and dependency: he became aggressive after loss and disappointment. The doctor said that the man's personality had not changed in Broadmoor and that the remedy for his offending was to provide understanding support for him in the outside world.

The second case was that of a young homicide offender who had been in Broadmoor for six years. He had no previous history of violence. Asked about the causes of the offending, the consultant said that although the patient had been having psychotherapy for 5 years 'the psychopathology hasn't been dug out, so we can't say'. As to what could be done; 'we need to get him to talk and to expose himself so that we can see what the problems are'.

Thirdly, the case of a sexual offender, whose consultant told us 'At the bottom of his offence is a sense of inferiority and lack of confidence due to his sexual inadequacy.' The patient had been in Broadmoor for well over a decade. His consultant told us that psychotherapy had not proved helpful: 'maturation of the personality' was the remedy now hoped for.

It will be seen that the problems of these patients, and the remedies proposed, tended to be described in generalized non-medical terms; and frequently the doctors were unable to point to any particular type of treatment that was appropriate for the condition said to cause the offending behaviour. The contrast with the psychotic men was striking. In some cases even after years in Broadmoor, consultants were unable to say what they thought had led the men to offend, let alone to suggest a remedy.

CONSULTANTS' VIEWS ON TREATMENT

The doctors were asked to rate the importance of medication, psychotherapy, and milieu therapy in the treatment of cases. For over 90 per cent of both psychotic and non-psychotic men, milieu therapy was rated as important. Medication, rated as important for over 90 per cent of the psychotic, was said for over 90 per cent of the non-psychotic to have no role to play. Psychotherapy was rated as important for most of the non-psychotic, but as Table 7.4 shows, there were sharp differences between the age groups. It was regarded as very important for 63 per cent of the youngest group but for only 7 per cent of the oldest.

We asked the consultants in each case whether they thought psychiatric treatment was currently needed. For the mentally ill men, the reply had been 'definitely yes' in over 90 per cent of the cases. For the non-psychotic, a very different picture emerged: only a minority, 38 per cent, were rated as definitely needing treatment. Among the oldest men only 10 per cent were so rated. Table 7.4 shows the position.

Uncertainty about the need for treatment was commonly expressed. For example, in the case of a sexual offender who had been in Broadmoor for a decade, the consultant answered our question about need for treatment by saying: 'Probably yes. But what treatment, I don't know.' That many of the oldest group were seen as doubtfully needing psychiatric treatment did not, obviously, mean that their consultants thought all was well with them. It simply reflected the unavailability of appropriate therapies. Because there was no treatment, and because it was not thought safe to release them, these men faced continuing preventive detention. The data showed that the patients who were held the longest were precisely those whom the consultants thought did not need psychiatric treatment.

We asked the consultants what they thought would be the relative importance of time and treatment in making each man fit for discharge. Men already being processed for release were excluded. The form of our question was as follows: 'In general, and looking to the future, what relative importance do you place on time and treatment in making this man suitable for discharge?' The doctors' replies are shown in Table 7.5; it excludes two cases in which the consultant thought that neither time nor treatment would render the patient fit for release.

The Table shows that consultants expected treatment to play a role greater than time in only 38 per cent of cases. For the psychotic men the proportion was 71 per cent. The Table also shows how differently the respective roles of time and treatment were rated for the different age groups. For a majority of the oldest men (57 per cent) time was seen as the main factor, whereas it was only so regarded for a quarter of the youngest men. The nature of the role that time was expected to play was also different. For the younger men, maturation was looked to as an ameliorating and beneficial process. For the older group, the benefits expected from the passage of time, the diminution of physical strength and sexual drive, were negative.

The consultants were asked what their future treatment plans were for the patients whose discharge was not proposed. For the majority, no specific treatment plan at all was suggested. Psychotherapy was the item most often mentioned, being suggested in 45 per cent of cases. Here again there were large differences between the age groups, psychotherapy being proposed for 63 per cent of the youngest men. It was evident that the doctors saw Broadmoor's function in relation to many of the non-psychotic men, particularly the older ones, as being largely one of care or asylum.

At the end of each interview with the consultants, we noted their views about the patient's treatability in the light of the time so far spent in Broadmoor. Although the men in the three age groups had been in Broadmoor for similar periods, the consultants' conclusions about their treatability was very different. The majority of the oldest men (62 per cent) were regarded as untreatable, compared with a quarter of the middle group and only 10 per cent of the youngest men ($x^2 = 21.43$ 2 df $p < 0.0001$). Of course only time will tell whether the optimism expressed on behalf of the younger men was justified.

SHOULD THE PATIENT HAVE BEEN ADMITTED?

In 60 per cent of cases the doctors said unequivocally that it had been right to admit the men, but they expressed reservations in the remaining 40 per cent. As would be expected, doubts were most often expressed about the admission of the older men. In response to our question 'should he have been admitted' the answer was 'definitely yes' for 76 per cent of the youngest men, 52 per cent for the middle group, and for only 44 per cent of the oldest men ($x^2 = 8.10$ 2 df $p < 0.02$).

The patients about whose admission the doctors expressed doubts fell into two categories. For about a third of them, it was thought that there had never been any need for maximum security, but in the majority of cases, and the older men were mostly included here, the reasons militating against admission were to do with the patient's untreatability. However, by no means all the patients who were regarded by their consultants as untreatable were put by them into the 'should not have been admitted' category. The doctors accepted that Broadmoor had a long-term care or custody function, and that some untreatable men were appropriately placed there. A number of the older recidivists were included in this group. The pattern of their repeated offences, often involving children or sadistic activities, meant that there was an obvious risk of further serious crime. The consultants were not happy that men for whom no treatment was available should be detained in hospital for indefinite preventive custody. But they believed, as did some of the men themselves, that since long-term detention was unavoidable, Broadmoor was providing it as humanely as was possible. Indeed, some of these men had come to Broadmoor from prison for this reason. For example, a sadistic offender with many previous convictions had been sentenced to life imprisonment by a judge who said that he should be freed 'only if there is a virtual certainty that never again will there be a repetition'. The prison medical authorities recognized that there was little prospect of treatment or change, and that he faced a lifetime in prison. It was thought that Broadmoor would provide him with a more humane environment, and he was transferred soon after his conviction. Many years later he was still there, and when we asked his consultant about treatment, the reply was 'It is really a question of time, till his sexual tendencies wane. . . . We have not done anything for him'. The provision of humane care for such a man should perhaps be prized as highly as the idea of treatment.

Another small group of patients about whose admission the consultants raised questions have already been referred to in Chapter 6, men who had been given determinate prison sentences and who, when the end of their sentences approached, were transferred to Broadmoor because the prison authorities considered them to be too dangerous to release. There were three such cases among the non-psychotic men in our sample, and in each one treatment had proved impossible because of the resentment and hostility which the men felt at having their sentences indefinitely extended by administrative means. A psychotic patient can if necessary be medicated against his will, but an offender who is not mentally ill cannot be forced to participate in psychotherapy.

In this respect these patients were like another small group, about whose admission the consultants also raised questions: the men who maintained that they were innocent. There were four such men in the non-psychotic group. A person who is convicted of offences that he denies cannot co-operate in treatment which is directed at exploring the reasons for his offending. Serious problems therefore ensue when such people are hospitalized. The hospital authorities have little choice but to assume that convictions are correct. But if they assume this, the patient's denials become a problem for his psychiatric treatment. Either his failure to admit to the offence may be seen as a delusion, or the authorities will feel that they cannot discharge him so long as he refuses to accept the need to examine and change his behaviour (Young & Hall, 1983). In the present context the point made by Broadmoor consultants was that people who persistently claim to be innocent should not, unless they are clearly mentally ill, be made subject to hospital orders.

If detention in a psychiatric hospital causes problems of treatment when the patient cannot be defined as psychotic and is unmotivated, they are as nothing compared with the difficulties which occur when the question of discharge arises for such men, as will be seen in the next chapter.

8 Discharge of non-psychotic men

SELECTION FOR DISCHARGE

This chapter is about the discharge decisions taken in respect of the 106 non-psychotic men. As with the psychotic patients, we interviewed the consultants about each man, and asked whether he was regarded as suitable for release. Of this population, 39 (37 per cent) were so regarded. We looked at how long the discharge nominees and other men had been in Broadmoor, and found that the groups did not differ on this variable, each having been in Broadmoor for 8 years on average.

The consultants were asked what level of security was required for the 67 men not being discharged: information was obtained for all but five of these. The consultants' replies showed that only a minority of these patients (23) were thought to need Broadmoor's maximum security. The doctors categorized 29 men as definitely not needing it and another 10 as probably not needing it. We therefore divided the population into three categories (1) the 39 men being discharged; (2) the 39 men who were not being discharged but were said not to require maximum security; and (3) the 23 men who were said to need maximum security.

In order to see what factors were associated with release decisions, we then compared the three groups on all the data we had collected. The significant findings that emerged clustered around two main areas: the nature of the admission offence; and the response of the patient to what Broadmoor had to offer.

The nature of the admission offence proved to be a factor of fundamental importance. The more serious it was, the less favourable were the patient's discharge prospects. When we rated the violence of the offences (Gunn & Robertson, 1976) we found that the maximum rating (offences causing serious injury or death) applied to a third of the discharge men, to half of the intermediate group, and to 78 per cent of those said to need maximum security ($x^2 = 11.83$ 2 df $p < 0.001$). Table 8.1 shows the offences. It is clear that homicide offenders are underrepresented in the discharge group and heavily over-represented among the group said to need maximum security. These findings are in complete contrast to the picture obtained for the mentally ill. For them, as was seen in Chapter 4, there was no association between gravity or violence of offence and adverse discharge status.

An indication of a different kind of factor that influenced the doctors came from our interviews with the men about their treatment experiences in Broadmoor. The men selected for release tended to be drawn from those who felt they had benefited from the hospital's psychotherapeutic facilities. As explained in the last chapter, we asked patients whether they thought being in Broadmoor had helped them at all. Although the great majority answered positively, there was a trend ($p < 0.10$) for the discharge group to do so more often. The difference may have been due to a halo effect, since the discharge group obviously had more reason to feel contented. However, a much larger and unexpected difference emerged when we asked the men who said that Broadmoor had helped them, to say in what way: two-thirds of the discharge group, as compared with 36 per cent of the intermediate group and 30 per cent of the maximum security group, volunteered the view that it had helped them to learn more about themselves ($\chi^2 = 8.45$ 2 df $p < 0.01$). The size of the difference suggests that the discharge group consisted in the main of men who had accepted the ethos (and terminology) of psychotherapy, and had succeeded in demonstrating to the doctors that they had acquired insight and self-understanding in the process. In other words, they had shown themselves motivated and responsive to what Broadmoor had to offer.

The importance of this to the consultants emerged from another set of data. We asked them to complete a motivation rating scale for all men not on the discharge list, and in the course of our interviews, asked them when they thought each of these men would be ready for release. When we compared the motivation scores of men whom the doctors expected to recommend for release within the next two years with the scores of those whose release was not imminent, we found that men in the former group were much more highly motivated: on a scale of 0–100, the average score for the former group (15 men) was 68 (s.d. 29), whereas for the other group (35 men) it was 36 (s.d. 26) ($t = 3.68$ $p < 0.0001$).

A cautionary note should perhaps be sounded in relation to these statistical findings. Although the data pointed to the importance of certain factors in the making of release decisions, none of these factors in themselves accounted for such decisions. For example, among men who had been in Broadmoor for at least five years, we found that those who told us they had gained in self-understanding were greatly overrepresented in the release group. Of the 27 men in that group, 21 (78 per cent) said they had made such gains, whereas among the 44 men not chosen for release only 14 (32 per cent) said that Broadmoor had helped them in this way. These figures certainly suggest that men who reported that they had gained insight had a good chance of being selected for release. Yet of 35 such men, 14 (40 per cent) fell into the non-discharge group. The point simply illustrates the complexity of the discharge decision-making process: consultants take so many factors into account, that no clear-cut and simple patterns emerge. There was only one exception to this: in no case within the present population, was a man proposed for discharge before he had spent less than three years in Broadmoor.

MEN SELECTED FOR DISCHARGE

The consultants told us that arrangements for discharge were to be made for 39 men: in most cases there was documentary evidence in the files to that effect, but in five cases it was only in the course of our interviews that the decision to release became known to us. Of these 39 men, almost half (18) were being proposed for discharge into the community. Places in local hospitals were proposed for the remainder, in nine cases in regional secure units.

(a) **How did men come to the doctors' attention?**

We asked the doctors how each man had come to their attention as being suitable for release. As with the mentally ill, the reason most commonly given was 'by a case conference'. This, together with other case reviews, accounted for almost half (47 per cent) of the discharge proposals; the cases all came from the lists of consultants who had been appointed within the previous two or three years. After routine case conferences, the most commonly mentioned 'surfacing' agent was the need to review the case for a Mental Health Review Tribunal. Consultants explicitly gave this reason for five men, but it seemed to have been relevant for several others. In the remaining cases, the doctors mentioned a variety of items, including the patient's own promptings, those of nurses or psychotherapists, and of course, the consultants' own observations. As with the mentally ill, the replies we received pointed clearly to the importance of effective routine reviewing procedures, in the absence of which other factors were bound to operate randomly.

(b) Changes in consultant

In some cases it was apparent that it was not changes in the patient, but changes in the view taken of him that had brought about the decision to discharge; reinterpretation of the case made discharge possible. For example, there were cases where the doctors thought that previous consultants had wrongly attributed offences to sadistic motivation whereas other, less sinister, explanations now seemed more appropriate. Another reinterpretative procedure was the retrospective attribution of the offence to a 'transient situational reaction'.

In about a fifth of discharge cases, a new doctor's review of the situation led to the release decision being made. Sometimes this was because the new doctor interpreted the case in a different light, but more often it was because he thought that enough progress had already been made to warrant discharge.

(c) Factors in the decision to discharge

In the early stages of our study, we had spent much time in trying to obtain from the doctors an appraisal of the factors that led them to consider that patients were ready to leave Broadmoor. As a result of this work, we drew up a list of items that the consultants had reported as being important, and we asked them to indicate for each of the 39 men on the discharge list, the role each of these factors had played. We also asked them, for each man, to list and rate any other factors that had been important in their decision-making. Presented with our list of items, the doctors in over 90 per cent of cases ticked three of them as having been important in their decision to discharge: a reduced risk of reoffending, a change in personality, and a change in behaviour in Broadmoor. It was rare for other reasons to be rated as important, but in a few cases the availability of new facilities, such as a medium secure unit, was said to have played a role, and in a couple of cases changes in the patient's circumstances were cited as relevant. In one of these the patient's wife had divorced him, and as he had been convicted of child battering, the divorce meant that he would not be returning to a home where children could be at risk.

For the three items which the consultants routinely ticked – change in behaviour, personality, and risk of reoffending – we tried to elicit further information. We asked when the changes concerned first occurred, but the doctors were seldom able to answer this. We also asked in what way behaviour and personality had changed. The replies about behaviour were too unspecific to make coding practicable. Replies about personality changes are discussed in the following section.

(d) Changes in personality and the concept of maturity

Table 8.2 shows what the consultants said when we asked them in what way the personalities of the discharge men had changed. The difficulty of eliciting hard information on this item will be apparent from the Table. The doctors ticked 'improvement in personality' as an important factor for virtually all their non-pyschotic discharge cases, just as they had ticked 'improvement in mental illness' as an important factor for nearly all the psychotic discharge men. When we asked them what had changed in the case of their mentally ill patients, they usually gave specific medical answers relating to the men's illnesses: for example we were told that men were now delusion-free or that their mood swings had stabilized. For patients who were not mentally ill, correspondingly specific replies were rarely forthcoming: the answers tended to be generalized, such as that the men were more mature or less impulsive. Replies of this kind must of course be viewed in context; the men were on average some 8 years older than they had been on admission.

Not only were the replies about the non-psychotic men less specific than those given about the ill, they also had less bearing upon the patient's criminal behaviour. As we saw in Chapter 3, the consultants said that for most of the mentally ill men there was a clear link between their crimes and their illnesses: they believed that the men's psychoses had led them to offend, and that effective control of the psychosis was therefore of major importance in relation to their dangerous behaviour. In contrast, as was seen in Chapter 7, the doctors did not as a rule link the crimes of the non-psychotic men with specific disorders for which particular remedies were available. So when these men were proposed for discharge on grounds such as increased maturity, the ways in which they were said to have changed did not have the direct relevance to their criminal behaviour that was cited for the mentally ill.

Given that men are held or released from Broadmoor on the grounds of what is said about their maturity, it is important to remember how vague and subjective an attribute this is. The following example provides an illustration. The patient was in his late twenties at the time of our research. He had been sent to Broadmoor at the age of 17 after a sexual assault on a child. His history was one of precocious sexual development combined with an inability to make appropriate relationships with the opposite sex. In Broadmoor his stay was uneventful, and he attended groups, social skills sessions, and a sex education course. Six years after his admission, he applied unsuccessfully to a Mental Health Review Tribunal for release. In his report to the tribunal, his consultant said he considered that: 'treatment in Broadmoor has so far made little impact on him'. The patient did not apply to a tribunal again, but eleven years after admission, now aged almost 30, his case was referred under the provisions of the 1983 Act. A case conference was held to determine what should be recommended, and the nursing report completed for this said: 'He is still very immature'. This could well have been presented as a reason for continued detention, but the consultant now in charge of the case considered that keeping the patient in Broadmoor would not help him to mature. He told us in our interview: 'He has been here 11 years . . . if we keep him, he will remain childish and immature'. He therefore recommended to the tribunal that the patient 'having improved and matured considerably' should be moved on. Thus a man described as very immature is nevertheless suggested for transfer on the grounds that he has matured considerably.

The case illustrates some of the difficulties surrounding the concept of maturation when it is made the touchstone of detention. The research raised many questions about it. What in these cases are the underlying assumptions being made about the relationship between immaturity and the potential for serious reoffending? Are they valid? If detention and release are to depend on a soft and undefined quality like maturity, how can arbitrariness be avoided? How far is the concept of maturity placed into service only because there are no hard objective criteria concerning the need to detain non-psychotic offenders in hospital? In this connection it was interesting to find that a disproportionate number of the small group selected for early discharge were men who suffered from neurotic disorders such as depressive mood. Improvement in these disorders was cited in their discharge recommendations as evidence that something had changed, and meant that reliance did not have to be placed on maturation alone.

(e) Sexual factors

In 19 of the discharge cases, consultants referred to changes they believed to have occurred in the patients' sexual functioning. For seven men reference was made to their increased sexual maturity, including the growth of self-confidence and discretion. In five cases it was believed that the patient's sexual orientation had changed from unacceptable to more acceptable objects, i.e. from juveniles to adults (only one of these patients had received specific treatment directed to this end). In other cases importance was attributed to the fact that men had come to terms with their homosexuality: this group included men who had been convicted of violent offences against women, and whose behaviour was seen as stemming from uncertainties of sexual identity. In a couple of cases consultants believed that the men's sadistic fantasies had receded.

In making their assessments about the men's sexual functioning, the consultants had generally not had recent recourse to psycho-physiological test results. Such testing facilities are available in Broadmoor, although for some of the time during our research they were not in use because of technical problems. Their usefulness for the purpose of establishing whether sexual offenders are safe to discharge is in any case limited. The tests provide information on what pictures and films excite the subject but throw no light on his wish or ability to control his sexual behaviour after release.

(f) Other factors

As stated earlier (Chapter 4), our attempt to elicit and describe the factors that went into the consultants' decision making was the most difficult and elusive part of our research, and there may be a danger that by trying to present our findings methodically, we are giving a misleading impression of what happens. In practice, the decision is based on the consultant's appreciation of the whole case; and though some of the ingredients in this can be identified or described, they do not add up to the whole. We have referred only to some of the most obvious of these ingredients, and those that can be counted in some way. There are many others, and their relative importance in each case is probably what in the end determines release decisions.

The following case illustrates this and also shows the relevance of factors such as the degree of personal rapport between patient and doctor. The man concerned was in his twenties and had been in Broadmoor for nine years. He had been admitted after a sexual offence, thought to be sadistically motivated. He had been in the care of his present consultant for just over a year and a recommendation for discharge had recently been made. The consultant told us he believed the patient's sadistic predilections had faded, that he had grown older, 'simmered down' and become more able to exert self-control. All this, which was certainly not the product of one 12-month period, had come to the consultant's notice because of a combination of events. A tribunal was one of these; a warmly favourable report from a psychotherapist was another. But the effect of the patient's personality on his new consultant was as important a factor as any other: 'As I got to know him, I felt he had genuine warmth and concern – I didn't feel he was dangerous'.

(g) Considerations of criminal justice

In each case we asked the consultant to say whether his decision to discharge had been influenced at all by the feeling that the patient's length of detention had already been commensurate with the gravity of his offence. With one exception, a case with unusual features, the answer was invariably in the negative.

(h) Internal delays in discharge proceedings

It was not surprising that the doctors proceeded with the utmost caution in coming to decisions about release. What did surprise us was that when they came to decisions it sometimes took them months or even years to act upon them. In one of these cases it was not until 15 months after a case-conference discharge decision had been made, that the consultant embarked on the necessary paper work. The patient meanwhile was aware that he was being detained because his consultant had not got round to writing a letter. He eventually became so anxious about the situation that he required treatment.

The implementation of better routine review procedures will, it is to be hoped, reduce the number of such cases. But it seemed to us that they stemmed partly from the institutional view of time. Time, as we learned, was regarded positively for its healing and maturing qualities, and in the present context it was even said to be useful to see how patients stood up to the stresses of discharge delays. Thus a year or two extra in Broadmoor was not necessarily seen as a serious loss of civil liberty, but more as a fairly unremarkable event that did no particular harm and might even prove to be useful.

PATIENTS NOT SELECTED FOR DISCHARGE

So far, we have been concerned with the decision to discharge patients. What of the opposite side of this coin – the decision to detain? Sixty-seven men came into this category.

In each case, we asked the doctors what would have to happen before they recommended release or transfer. In a third of cases, they said that the men's sexual problems would have to be resolved or their sexual drive to have waned, in over a quarter (28 per cent) that the patient would have to be more mature, and in just under a quarter (24 per cent) they said that the patient's psychopathology would need to be clarified or dealt with.

The above items were those most frequently cited; they were not mutually exclusive. Other factors also mentioned were the need to be more open and to discuss problems (19 per cent), to cope better with stress (16 per cent), to demonstrate a period of stable behaviour (8 per cent), and to show motivation for change (7 per cent). All these items were signals which the doctors said they would need to see before discharge could be envisaged, but clearly none of them would in itself have been sufficient to lead to a man's release. For example, although doctors referred to the need for the patient to be more open in discussing his problems, it is obvious that this in itself would never be a reason for release; it could pave the way for a discharge decision only if the results of greater openness gave the consultant confidence about the man's future conduct. Such a feeling of confidence stems from an almost unlimited number of possible sources.

Two other prerequisites for release remain to be mentioned. In about 10 per cent of cases, the doctors raised an item that was independent of the patient's condition – for example the existence of an obvious potential victim. In two cases the consultants said that they themselves constituted a bar to discharge in that they did not know the patient well enough to have an opinion on the question.

Finally, in about 6 per cent of cases, the need to find a suitable placement was mentioned as a prerequisite for release. Usually the men concerned had special needs, such as institutionalized patients who needed looking after but did not require a hospital bed. However, in another sense, it could be said that the majority of the men were being held in Broadmoor for lack of a suitable alternative placement. For as we have already seen, when we asked the consultants whether the 67 men required maximum security, it was only in respect of a minority, 23 men, that the answer was 'yes'.

REASONS FOR DETENTION

After we had seen the patient, his doctor, and the hospital records, we tried to summarize in each case the basic reasons why the patient was held. The picture that emerged, when all 67 summaries were put together, rather resembled the cards available in paint shops, on which various shades of household paints are displayed. The darkest shades represented the cases where there were very obvious and powerful reasons for the patient's detention, and at the other extreme were the very light shades representing cases that could really not be distinguished from those in the discharge group.

(a) Preventive detention

In the darkest areas of our shade card, were men who had been convicted on more than one occasion of major offences – homicide, sexual assaults, arson. Essentially they were being held for preventive purposes in the interests of public safety.

The patients in this group presented a variety of problems to their consultants, but agonizing over the pros and cons of their release was not one of them. The men had repeatedly committed life-threatening offences, were still fit enough to do so, and Broadmoor had provided no remedy. Not all of these men were rated as needing maximum security, but the doctors believed that no other hospital would take offenders who were not mentally ill and who required long-term detention. In most of these cases the motivation was sexual and, because of the nature of the problem or because of the man's age, the doctors had no effective treatment to suggest. The most that was hoped for was that eventually, with time, there would be a diminution of strength and sexual drive. Indefinite detention until then was the only prospect.

(b) 'Recent' arrivals

There was an assumption on the part of most consultants that non-psychotic offenders convicted of serious crimes are bound as a rule to spend some years in Broadmoor before they can be proposed for discharge. The assumption is institutionalized to some extent, in that it was unusual for newly admitted patients to start specific treatments like individual psychotherapy or behaviour modification within the first years of their admission. For example, a young man who was admitted after a very violent sexual offence had been in Broadmoor for more than two years when he was seen for our research. He had not yet taken part in any specific treatment programmes, although from the assessments made it was clear that he needed a great deal of help in coming to terms with his sexuality. His doctor did not suggest that there was any urgency about starting with this work, apparently because it was assumed that the patient would be in Broadmoor for a number of years in any case. The patient expressed frustration about the lack of any specific treatment: 'They say that being here is treatment. But if that's true, I could sit around for twenty years . . . I want positive treatment to help me get out'. The truth of his words was demonstrated when he applied for release to a Mental Health Review Tribunal. His consultant opposed it on the grounds that he had serious sexual problems which had not yet been treated, and on which 'there is still much work to be done'. The patient was in a Catch 22 situation: he could only leave Broadmoor if he could demonstrate a successful response to treatment, but treatment for his specific problems was not forthcoming, since by Broadmoor standards his settling-in period was scarcely over.

(c) Waiting for evidence of change

We noted in Chapter 6 the importance that doctors attributed to the role of time in making men ready for discharge. As one consultant put it: 'We've got to keep trying different things, till Time the Great Healer does the trick'.

There are many problems about waiting for the Great Healer to do the trick, and one of them is the difficulty of identifying the moment when it has been accomplished. In men who are not mentally ill, the changes being looked for are undefined, and their recognition, as the doctors were ready enough to tell us, is not a scientific process. As one consultant said when we asked him why his patient was still being held: 'It's a gut feeling . . . I would find it hard to justify persuasively to an outside psychiatrist'.

Not only were the changes being looked for unspecific, but their relationship to treatment was also far from clear. Most of the men had taken part in psychotherapy, but however favourable their response, it did not follow that their doctors would want to discharge them. Whereas for the psychotic men good response to treatment was seen as having definite relevance to the likelihood of reoffending, for the non-psychotic, response to psychotherapy was not seen in the same way. For example, a man with a serious record of previous convictions had been in the hospital for six years. He was reported to be making good use of psychotherapy, but his doctor pointed out that there could be no certainty as to what bearing this actually had on the likelihood of his reoffending: the number of previous convictions was seen as having a more certain relevance, and remained a bar to his release.

The group of patients considered 'not yet ready' thus consisted of men about whom the consultants felt uneasy. Sometimes, as in the case just cited, however co-operative the patients were, it was clearly their records that militated against release. Sometimes the doctors referred to lack of trust, and doubts about the patient's veracity. 'I don't trust him, so I don't play' was how the consultant put it for one man. Sometimes patients refused to take part in treatment or offered only token co-operation, so their unfitness for release was thought to be clear. More often though, they were said to have matured, simmered down, or changed to some extent, but not sufficiently. But how does a doctor know when his patient has simmered down enough? How does he define maturity? How is it to be tested in a segregated maximum security hospital? Above all, what does he assume the relationship to be between 'immaturity' and the potential for dangerous reoffending? To these questions, which lay at the heart of our inquiries, no objective or satisfactory answers were forth-coming. Again, the contrast with the psychotic men was apparent. For them, hard medical criteria were available to establish whether their illnesses had improved, and though other factors also affected discharge decisions in their case, the medical contribution was crucial and obviously valid.

(d) **Transient situational reactions**

In Chapter 6 we discussed some of the issues that surround the hospitalization of certain offenders who have no history of delin-quent behaviour. Once a patient of this kind is admitted to Broad-moor he becomes subject to the assumption that underlies the detention and discharge process for all such offenders – namely that he may repeat his crime, unless there is evidence to the contrary. But for men who are not mentally ill, who have no previous offences, and whose crimes were committed in transient or unusual circumstances, it is extremely difficult for the doctors to know what evidence of change would be material to the risk of reoffending. Nevertheless, until a consultant states that he has found such evidence, detention will continue. The following case illustrates the issues concerned. The patient had no previous offences and had, quite unexpectedly, killed a member of his family. His five-year stay in Broadmoor had been uneventful. An experienced forensic psychiatrist came to examine him for a Mental Health Review Tribunal hearing and reported as follows: 'The events arose within the family (of origin) which was the seat of his conflicts. It would be inconceivable almost for a similar set of circumstances to prevail again and he must represent an insignificant risk to the community'. Nevertheless, the Broadmoor doctor recommended continued detention on the grounds that it would serve to determine whether the patient's well-established stability could be sustained in the face of adversity and disappoint-ment.

In some of these cases it was diagnostic reinterpretation, from psychopathic disorder to transient situational reaction, that eventually paved the way to the patient's discharge. The following case exemplified the process. The man had no previous convictions or history of mental disorder, but was sent to Broadmoor as a restricted patient following a series of fire-raising offences. Three years after his admission, he applied for release to a Tribunal. His application was supported by the report of a psychiatrist who came to examine him. He wrote that the offences were committed in a transient situational reaction and that he did not consider the patient had ever come within the legal definition of psychopathic disorder. The tribunal accepted this doctor's opinion and advised the Home Secretary (under the 1959 Act Tribunals could not discharge restricted patients) that the patient should be released. His consultant did not oppose this course and discharge was proceeded with.

(e) Time for crime

There are factors in the detention process for non-psychotic offenders that give it something of the appearance of a tariff system. With the psychopathic disorder patient, the attribution of the disorder, as we saw in Chapter 6, stems essentially from the offences committed. So if the gravity of the disorder is defined in terms of what the offender has done, it is not surprising to find that the more serious the crime the longer the detention. There are also tariff connotations in the concept of maturation as a cure, or antidote, for psychopathy. If maturation is the remedy, then time will be regarded as the dosage.

The consultants resisted the suggestion that they ever detained non-psychotic offenders on a 'time for crime' principle, or because they took it for granted that the Home Office would not agree to discharge. But there were cases in our sample where it seemed that the hidden reason of 'he has not served long enough' was relevant to the patient's continued detention. The following case is an illustration. The patient was a young man who had been in Broadmoor for four years, after an unusually callous homicide. He had made an excellent response to psychotherapy, and in our interview his consultant rated him as substantially improved, highly motivated, needing no security, and constituting no danger to others. Nevertheless the doctor was not recommending release, saying that he thought the patient should stay in Broadmoor longer in order to complete his treatment and to 'mature' further. The words of a man quoted earlier seemed apposite: 'They don't ever have to say you haven't done enough time . . . they can always find a psychiatric reason.'

(f) **Knowing and understanding the patient**

We have already noted the obvious fact that for a patient to be recommended for discharge his doctor had to know him sufficiently well. For a variety of reasons, including staff changes, some patients were not well known to their consultants and this effectively was the reason for their detention. Similarly, continued detention faced men whose offences their doctors could not explain. If a consultant cannot understand what caused his patient to behave dangerously, he is unlikely to be able to say that the propensity to behave in this way has waned, and hence to recommend release. Difficulties in understanding the patient's behaviour when it cannot be explained in medical, mental illness terms, can thus prevent discharge for many years. The following case illustrates the process. The man had been admitted at the age of 17, after an assault offence which carried a maximum penalty of five years' imprisonment. He had attacked a woman unknown to him and had done the same thing before. In Broadmoor, he produced no evidence of psychiatric disorder. He was well-behaved but unforthcoming. He had been there for 10 years by the time of our research, and had been in the care of several consultants. We asked the current one what would have to happen before discharge was recommended. The reply was 'We would need to have an understanding of why he committed the offences.' Because he felt unable to explain and therefore to predict the man's behaviour, let alone to suggest treatment for it, the consultant, like his predecessors, was not prepared to take the risk of discharging him.

(g) **Patients unwilling to leave**

Over a quarter of the patients not being discharged told us that they did not yet feel ready to go. Sometimes this was because they wanted to finish treatment courses they were on, (and in these cases, the consultants also thought discharge was inappropriate), but more often it was because they had anxieties about how they would cope in the outside world. Some of these men threatened to commit serious offences if they were released, but it was more common for them to tell their doctors that they did not trust themselves not to re-offend. Either way, they effectively deprived the doctors of the power to discharge.

(h) **Limitations of snapshot research**

The preceding account represents a snapshot of reasons for detention taken at the time of the research, but it has to be remembered that over the long periods of time these patients were held, a variety of reasons had often been relevant. A longitudinal survey of the reasons that patients were held would present a different picture from our snapshot.

The following case illustrates this point. The patient had been admitted in his thirties for a sexual offence which he denied. His first doctor noted that treatment was impossible since the patient denied the offence, and that the denial ruled out the possibility of release. 'I cannot see my way to making a recommendation for discharge until he is an old man.' For some years, therefore, the patient's denials were the primary reason for his detention. When subsequently a different doctor began to moot the possibility of releasing him, the patient was unwilling to leave: a new reason for detention thus came into operation. After this, and under another consultant, a third factor came into play: the patient had disappeared into the wood-work. When, fourteen years after admission (in the course of which he had one year's group therapy), a new consultant reviewed the case, he recommended discharge. The patient was now thought to be trustworthy: 'a matured middle-aged man who had lost the wild ways of his younger days'.

PATIENTS' VIEWS

We asked the men whether they thought there was anything they themselves could do to secure their release. Good behaviour ('keeping your nose clean') and co-operation with treatment were the two things routinely mentioned. The former was simply a negative requirement: no one believed that good behaviour got people out of Broadmoor, but all recognized that bad behaviour would delay release. Co-operation with treatment was a more positive matter, and for some men it had difficult implications. 'They want to change you . . . the system wants you to agree with everything the doctor says about treatment . . . I want to get out of Broadmoor, so I agree with them. I play the game, and deny myself the luxury of dis-agreement.'

When we asked patients what would have to happen before they could be discharged, the first point commonly made was: 'you have to serve the right length of time'. Of course it was recognized that time, like good behaviour, was not itself a sufficient reason for discharge: 'They can't let people out just because they have been in Broadmoor for 15 years'. But time was seen as an essential prerequisite. When we asked a sexual homicide offender who was getting excellent reports from his therapist, what would have to happen before he was released he replied: 'Time must pass . . . six years isn't enough for what I've done'.

THE NEED FOR MAXIMUM SECURITY

As already reported, we asked the consultant in each case to say whether he thought the patient's security needs could be met elsewhere than in a special hospital. Information was provided for 101 of our 106 men. Thirty-nine men (39 per cent) were being processed for discharge, so by definition were regarded as needing less than maximum security. Another ten men (10 per cent) were described as probably suitable for less than maximum security and 29 men (29 per cent) were described as certainly not needing maximum security. Thus of the 101 men representing virtually the whole of Broadmoor's resident population of non-psychotic men, only 23 (23 per cent) were judged by their consultants to need maximum security.

We asked the doctors why no attempts had been made to transfer the men they described as needing less than maximum security. We were told that regional secure units would not take offenders who were not mentally ill and who needed long-term care. Local hospitals, even where they were thought to be adequate for the patient's security needs, were ruled out for the same reason.

We examined the data to see what features, if any, characterized the men who were said not to need maximum security but who were not to be discharged. Two findings were of interest. First, there was a trend for these men more often than others to have shown evidence of profound sexual problems. A number of them were burdened with a sexuality that made them a danger to others, and in some ways their predicament and need for suitable long-term asylum was similar to that of the chronically mentally ill men.

The second point of interest was that the men described as not needing maximum security but not being discharged, included a disproportionate number whom the doctors were thinking of discharging within the next two years (40 per cent, as compared with 9 per cent of the maximum security group, $p < 0.02$). Together, the two findings suggest that it would be valuable if Broadmoor had some sort of less secure facility to cater both for certain long-term patients, and for those whose discharge is under consideration.

DISCHARGE PATTERNS IN AN ADMISSION COHORT

Our sample consisted of Broadmoor's resident non-psychotic population; released men were excluded. To get a full picture of how release procedures work, and in particular of how long patients are detained, an admission cohort needs to be studied. To some extent we were able to do this by using data on admissions provided by the Special Hospitals Research Unit. This study is reported in detail elsewhere (Dell *et al.* 1987) and only the salient findings relating to length of stay will be referred to here.

The sample consisted of men on restriction orders admitted to Broadmoor from the courts between 1972 and 1974: of these 76 had a legal classification of psychopathic disorder and 111 had a legal classification of mental illness. Data were provided about the men's psychiatric and criminal history, their admission offences, and their dates of discharge, if any, by the end of December 1982. Over half (53 per cent) of the PD men were detained for 8 years or more, as compared with 42 per cent of the mentally ill. Thirty per cent of the mentally ill offenders were released after spending three years or less in Broadmoor; only 10 per cent of the PD group were released within this period ($p < 0.01$).

For men admitted under the PD classification, the factors relating to length of stay were not those connected with the patients' characteristics, such as their psychiatric history, but those connected with their offences at admission. Men who had not injured anybody accounted for the majority of short-stay cases, whereas those convicted of violent or sexual offences, particularly against strangers, made up the majority of the long-stay group. The figures showing how long different kinds of offenders were detained reflected the seriousness of their offences.

Similar analyses were carried out for restricted offenders admitted under the mental illness classification. For them, no significant association emerged between length of detention and gravity of offence committed: it was measures reflecting the severity and chronicity of illness which were linked to length of stay. Whereas the great majority of PD men convicted of homicide or wounding offences fell into the long-stay group, among the mentally ill such offenders were as likely to fall into the shortest stay as into the longest stay category.

DISCUSSION

Our study of Broadmoor's resident population and our admission cohort both showed the discharge decisions for PD offenders to be related to the gravity of their crimes. From our interviews with the patients, we were left in no doubt about their recognition of this basic reality. Nevertheless, the official ethos of the hospital is that 'time for crime' plays no role in the discharge process, and that patients are released as soon as they are 'better'. The problem, of course, is that for those who do not suffer from identifiable illnesses, hard objective criteria for being better are absent. The situation, and its attendant official 'double speak' was the cause of much cynicism among these patients.

The issue has recently been aired by a doctor who was working in the hospital at the time of our research (Grounds, 1987a): 'Although in principle the hospital order is imposed to enable treatment, and carries an implicit promise of release when the patient's condition is improved . . . the reality is more complicated. In practice covert tariff considerations do apply for the psychopathic offender under a hospital order in a special hospital. Carefully veiled intimations of this occasionally emanate from the Home Office, and tariff considerations also colour our own clinical judgement although sometimes this is masked by our psychiatric vocabulary. I commonly hear from nursing colleagues the view that a particular individual who has committed a grave offence has a long time to do or that his problems will take a long time to resolve – the two notions are interchangeable'.

The absence of medical discharge criteria emerged as a constant theme when we interviewed the consultants and asked them to explain what had led them to make detention or discharge decisions for the non-psychotic man. In making their discharge judgements about men who were not mentally ill, the doctors were essentially making predictions about reoffending. They were guided by a variety of non-medical considerations, including both what they called their gut feelings, and the basic criminological principles of reconviction, one of which is that young men generally calm down as they get older.

9 Discussion: non-psychotic men

PSYCHOPATHIC DISORDER—A COMMENT ON ITS DERIVATION

Personality is a term used to describe the whole self, the persona or total being. In the history of psychology, it has its roots in philosophy and in religious ideas about the soul. The behaviourist reductionist tradition within psychology has shown scant interest in such notions but it is from this latter branch of psychology that the functional concept of personality disorder has gained its credibility. In psychiatry, it is the dynamic tradition which has concerned itself with personality and with personal development and awareness. The self and its vissicitudes have been the province of the analytically inclined. As with psychology, the religious and philosophical over-tones are very strong. The fact that the concept of personality disorder can draw from both traditions lies at the heart of many of the problems and frustrations voiced by both staff and patients when we asked them about treatment in Broadmoor.

As a rule, people who are not suffering from a mental illness are only admitted to Broadmoor Hospital after they have committed a serious offence. This offence may be regarded as a piece of behaviour which allows admission to take place. Sometimes the offence represents part of an established pattern of violent behaviour but more often, the behaviour which gains admission is represented by one act. The concept of personality derived from the behavioural tradition is invoked – 'He is dangerous/personality disordered because he did something'. Once in Broadmoor, the behaviour which brought about the admission is (with very few exceptions) not repeated, but this does not bring about discharge. Instead, the broader concepts of the nature of personality, rooted in the dynamic view of man are referred to, allowing continued detention to take place and bringing with them cynicism and resentment on the part of many who are so detained. What the man did gains his admission to Broadmoor, yet once he is there he is likely to be told that it is not what he did but what he is that is keeping him indefinitely detained. The mentally ill are defined by that from which they suffer, and their illnesses may remit or may respond to treatment. On the other hand, the so-called psychopath will continue to be defined by that which he has done and this makes for a crucial difference between the two groups. What is done cannot be undone.

CHANGES IN THE 1983 MENTAL HEALTH ACT

Our study was conducted when the 1959 Act and its definition of psychopathic disorder was in force. Have the changes in the 1983 Act, particularly those concerning alleviation and deterioration, made any difference?

The new Act made three main changes. The first impinges only on civil patients: there is no longer any age restriction on the detention of psychopathic disorder patients under civil procedures. The change has not affected psychiatric practice; the detention of civil PD patients remains extremely rare (see below). A second change has been an extension of the types of behaviour which by themselves are not to be regarded as proof of the existence of disorder of mind. The 1959 Act stated that a person could not be defined as suffering from mental disorder 'by reason only of promiscuity or other immoral conduct'; the 1983 Act added 'sexual deviance or dependency on alcohol or drugs'. This change has made no practical difference.

The third change is that the new Act has dropped the clause 'and requires or is susceptible to treatment' and in its place section 3(2)(b) of the 1983 Act stipulates that for a treatment order to be made in respect of someone who is psychopathically disordered, it must be stated that 'such treatment is likely to alleviate or prevent a deterioration in his condition'. The fact that this change has had little practical significance emerged clearly from our research. All the PD men in our study had been admitted under the 1959 Act and at the end of our interviews with their consultants we asked them to say from their knowledge of each case, whether on admission the offender had (a) fulfilled the treatability conditions of the 1959 Act, and (b) fulfilled the conditions about alleviation and deterioration in the 1983 Act.

There were no cases in which the consultants thought that a patient had satisfied the requirements of one of these Acts but not the other. We consequently stopped asking the question about half way through the study.

DIFFERENCES BETWEEN
PSYCHOTIC AND NON PSYCHOTIC MEN

The two types of patient for whom Broadmoor cares differ fundamentally from each other, and differ fundamentally in the way the doctors regard them. The first point is obvious enough clinically, and emerged clearly in statistical terms when we examined the data collected on patient admissions by the Special Hospitals Case Register. Tables 9.1 and 9.2 show, for restricted hospital order cases, the main differences between men admitted in the legal categories of mental illness and psychopathic disorder. Offenders admitted in the latter category came to Broadmoor younger (on average they were aged 25 on admission, whereas the mentally ill group were 34), and after considerably more disrupted careers. They had experienced more family disturbance and more institutional care and were much more likely to have been before the juvenile courts. Though much younger, they had also been convicted and imprisoned by the adult courts more often, particularly for violent and sexual offences. Only a minority of them had ever been psychiatric in-patients, as compared with the great majority of the mentally ill.

Our research in Broadmoor revealed major differences between the attitudes of the two groups. The non-psychotic recognized (Chapter 7) that they would have gone to prison if they had not come to Broadmoor, and a majority felt that being in Broadmoor was a more comfortable, relaxed, and better-paid way of doing their time; they also appreciated the treatment opportunities. The psychotic men, however, knew that for them the alternative to Broadmoor would have been a local hospital, and the majority felt (Chapter 3) that they were worse off in Broadmoor, where they were held longer and were cut off from the world in conditions of massive security. Thus, when we asked general questions about the men's opinions, we found that the non-psychotic patients were markedly more pro-Broadmoor than the psychotic.

In one important respect, however, the attitudes of the two groups crossed over. When we asked whether they felt satisfied with the treatment they had received, a majority of the psychotic men (54 per cent) but only a third of the others (34 per cent) answered in the affirmative. The non-psychotic men often expressed unhappiness at the lack of programmes directed towards their specific needs, feeling that they had come to Broadmoor on a false prospectus – one that promised them treatments that were not forthcoming.

It was clear from our research that the Broadmoor doctors regarded the two groups of patients in totally different ways. About the need for the mentally ill men to be in hospital (though not necessarily in Broadmoor) and about the kind of treatment they required, the consultants rarely expressed doubts. About the need for the non-psychotic men to be receiving hospital care, however, uncertainty was commonly voiced. From the replies to our questions and schedules, (Chapter 7) we learned how doubtful the consultants often were as to whether and what psychiatric treatment was appropriate for these men. In response to our question 'Does this man need psychiatric treatment?' the answer was 'definitely yes' for 92 per cent of the psychotic, but for only 38 per cent of the non-psychotic.

When we came to look at treatment, the differences between the groups reflected the doctors' views. We noted what specific treatment each man was given in addition to the one that all of them received – that of being in Broadmoor. At the time of our research, the majority of the psychotic men were being treated by medication, but the majority of the non-psychotic men were not receiving any treatment other than being in Broadmoor (Chapter 7). They had usually participated in psychotherapy at some point in their stay, but for most of the time – on average for at least two-thirds of the eight years for which they were detained – being in Broadmoor was the only treatment they had.

Contrasts between how the consultants viewed the two groups of patients emerged at many other points in the research. When we asked them to say what they thought had caused each man to offend, the answers for the mentally ill men were, as a rule, put concisely into a medical framework. The doctors believed that it was their psychoses that had led these men to commit their crimes, and that effective treatment (with medication) was therefore the remedy for their dangerousness. In the case of the non-psychotic men such clear cut medical answers were not forthcoming. Uncertainty about causes and remedies was frequently expressed, and also about the relevance of available treatments to the risk of reoffending. Moreover, it was not unusual, even after years of detention, for the doctors to confess that they were ignorant about the causes of the offending behaviour. 'We can't say what the problems are . . . the psychopathology hasn't been dug out.' These were one consultant's words when we asked him about a man who had been in Broadmoor for six years.

The fact that consultants were liable to confess ignorance as to what treatment would help their non-psychotic patients reflected the way that these men had been admitted. It was their crimes, not their psychiatric needs, that had initiated the medical committal. But for their offences, doctors would not have said they needed to be in hospital. It is therefore not surprising that the Broadmoor doctors were sometimes at a loss to know what treatment was appropriate. Again the contrast with the mentally ill men was evident. Most of the latter were suffering from disorders that would have led them into psychiatric care even if they had not offended: their illnesses, not their offences, were the focal point in their hospitalization.

All these matters were reflected in the approach of the consultants to the discharge of their patients. For the mentally ill men, the factor of primary importance was the course of the illness. The men were seen as having offended as a result of their illnesses and, if they got better, transfer to lesser security became a possibility, whatever the offences. In the appraisal of mental state for this purpose – for example, in monitoring the intensity of delusions, or the response to medication – the doctors were expert. But for the non-psychotic, the assessment of improvement was far more problematical. There was usually no obvious mental disturbance to monitor, no clear-cut response to treatment to assess, nor any hard medical criteria by which the men's readiness for release could be judged. Greater maturity, or 'simmering down' were the factors most commonly mentioned by the doctors in relation to their discharge decisions. However, the statistics showed (Chapter 8) that discharge decisions for the non-psychotic men were closely related to the seriousness of their offences. The graver these were, the less likely discharge became. For the psychotic offenders there was no such relationship.

Thus at every turn, admission, treatment, and discharge, there was evidence of the profound differences between the two groups of patients, and the way they were regarded. The psychotic were seen as falling unequivocally into the psychiatric province, a clearly medical framework existing for the understanding of their offences, their treatment, and their release. For the non-psychotic men no such medical framework or understanding was in evidence. It was clear that Broadmoor's consultants shared many of the same doubts and uncertainties about the concept of psychopathic disorder and its treatment that Lewis (1974), Walker and McCabe (1973) and the Butler Committee (Home Office, DHSS, 1975) have chronicled (see Chapter 6).

TREATABILITY

The conclusion of the Butler Committee about the concept was as follows: 'Psychopathic disorder is no longer a useful or meaningful concept. . . . The class of persons to whom the term relates is not a single category, identifiable by any medical, biological or psychological criteria'. (para 5.23) Some ten years later, Chiswick and others (1984), describing their research on the prosecution process in Scotland, showed that the unwillingness of psychiatrists to accept psychopathically disordered offenders for treatment stemmed not only from their therapeutic pessimism, but from basic doubt about the psychiatric validity of the concept. Chiswick and co-workers also found psychiatrists strong in support of the Butler Committee's conclusion on treatability. The Committee had reported 'The great weight of evidence presented to us tends to support the conclusion that psychopaths are not in general treatable, at least in medical terms' (para 5.34). Chiswick and co-workers reported 'an almost universal view expressed by psychiatrists . . . that this form of mental disorder could no longer be considered amenable to psychiatric treatment'.

More recently Grounds (1987b) after some years' work at Broad-moor, published his conclusions about the non-psychotic offender patient: 'In the case of "psychopathic disorder" if there was evidence – first, that we were dealing with a well-defined disorder; secondly, that we had specific treatments for it; thirdly, that we knew who was likely to benefit from treatment; fourthly, that such treatments were effective in producing psychological change; and fifthly, that psychological change implied a reduced risk of dangerous behaviour; then there would be justification for preferring to deal with such offenders by indeterminate treatment orders rather than tariff sentencing. But for the "psychopath" the evidence of each step is lacking and is insufficient to support the practice of recommending hospital orders for them.'

The body of medical opinion quoted above is reflected in current psychiatric practice. If people suffering from psychopathic disorder really needed compulsory treatment as in-patients, they would be found in ordinary psychiatric hospitals. They are in fact few and far between. On 31.12.84, there were no more than five men detained in local hospitals with the category of psychopathic disorder under section 3 of the 1983 Act (DHSS, verbal communication). Detained PD patients are to be found mainly in the special hospitals and virtually all are offenders. None of the men in our study who after their offences were admitted to Broadmoor under the PD category had ever been detained as psychopathically disordered under the civil provisions of the 1959 Mental Health Act, though all had been eligible for such detention when they were under 21. It gives some measure of the abnormal nature of Broadmoor's population, to point out that it contains thousands of times the number of detained psychopathic patients that one would expect to find in a local psychiatric hospital (see Chapter 1).

In our opinion, the problems which occur in the treatment and discharge of offenders who are not mentally ill have their roots in the procedures which bring these men into hospital. The Mental Health Act requires a psychopathic detainee to have 'a persistent disorder or disability of mind'. In practice the crime is regarded as proof of this disorder: had it not been committed, psychiatrists would not have considered the man to need hospital care. However, once the crime has occurred, the doctors – if they think a hospital disposal desirable – are forced into one of two procedures. Either they will cite little other than the offence as evidence of the disorder; this, as we saw in Chapter 6, was commonly done. Alternatively, in an attempt to avoid this tautological approach, they will cite as proof of disorder a number of features (for example, poor family or personal relationships) that existed before the offence, but would never of themselves have caused the doctor to recommend hospitalization. Whichever approach is used, it is the offence, and not any 'disorder of mind' which brings the non-mentally-ill patient into hospital. The whole process can be regarded as a useful fiction. We understand its utility, but question the need for the fiction, as it carries with it the potential for abuse, for injustice, and for waste of both time and precious resources. We discuss below ways in which we believe the system could be reformed.

The procedures we are questioning have their basis in the process whereby the offender's behaviour is conceived of as a medical matter. All aspects of human functioning have biological correlates and hence, if wished, may be regarded as having a biological/medical component. The question is whether what is known medically about the people described as psychopaths justifies the use of the concept to sentence them to indeterminate hospital detention rather than to prison. In this regard, we repeat a question raised earlier. What incremental validity does the medical concept possess? What, other than its technical language, does the medical speciality of psychiatry contribute to our knowledge and understanding of personality? Because it represents another way of looking at personality, it probably adds a little to that understanding – but no more than a little. In this, the medical contribution to personality theory is not more or less than that contributed by many other disciplines. The difference is that the law affords the medical attribution of psychopathic disorder considerable power – the power to detain in maximum security for life. Does it merit this distinction? What we learned from our research brought us to the conclusion that it does not.

In ordinary life we present ourselves to doctors in order to seek expert advice – usually to be relieved of some pain or discomfort. In two types of medical practice, emergency work and psychiatry, this paradigm may not hold good, and the patient may be an unwilling or unconscious participant in the social exchange that takes place with the doctor. In psychiatry, this situation arises when the patient is so mentally disordered and his powers of reason so impaired by psychotic illness that he is regarded as being 'out of his mind'.

Men held under the label of psychopathic disorder are not out of their minds, yet are subjected to compulsory medical treatment. What is desired as a rule is that they should be made to behave better, and not to reoffend, but the official reason for their detention is their supposed need to be made better in the medical sense. Hence terms with medical connotations are employed in the legal framework, like 'suffering from', 'condition', 'disorder', and 'treatment'; terms which create the illusion that cure is being offered for illness. We see no reason why this fiction need be maintained and believe that the help which is available in psychiatry and clinical psychology could be offered these same men without the creation of a legal framework which is so detached from reality. Before we turn to these issues, however, another one must first be addressed.

WHAT IS THE PROBLEM?

It has been put to us that the problems we found in Broadmoor in relation to the non-psychotic men arose from the deficiencies of the hospital, rather than from deficiencies in the legal and conceptual framework under which they were held. Several points have been made to us in this connection. The first implies that Broadmoor's doctors are a race apart and that it is they and not the legal procedures that need to be reformed. Our research did not point to such a conclusion. At the time of our study, Broadmoor consultants comprised a varied group in terms both of age and experience. Some had been in the hospital for many years, but some had come more recently from other fields, including general psychiatry, and there were also those who held joint appointments as senior lecturers at the Institute of Psychiatry. It would not be possible to maintain that the Broadmoor consultants were a homogeneous group, practising methods and standards of psychiatry different from those regarded as appropriate by other members of the profession.

Nevertheless, it has been suggested to us that the frankness of the doctors in expressing the dilemma they faced in treating non-ill patients, and the lack of specific treatments given to many of these men, reflected the peculiarities of the hospital and its staff rather than the peculiarities of the task with which they are charged. Again, our research did not support this conclusion. As we have seen, psychiatrists have for decades questioned the concept of psychopathic disorder, the treatability of psychopaths, and the appropriateness of the medical model for people who are not ill but show behavioural or personality problems. The dearth of treatments for such men and lack of medical confidence in their treatability are not peculiar to Broadmoor hospital. It should also be said in this connection that Broadmoor does not function in a vacuum; psychiatrists and psychologists from outside visit and examine its patients, review their treatment, and prepare reports on them for various purposes. This happens routinely when Broadmoor consultants invite their National Health Service counterparts to visit patients whose transfer is being sought, or when patients themselves commission reports for the purpose of presenting them to Mental Health Review Tribunals. It was apparent from reading such reports, which are made by a wide variety of forensic and other psychiatrists, that the doctors who came to examine non-psychotic patients did not as a rule have treatment recommendations to make. Sometimes, where no specific treatments had recently been undertaken, the reports might suggest that psychotherapy should be attempted. But there was no suggestion in them that a fund of alternative treatment strategies existed for these men which the Broadmoor doctors were neglecting to apply.

A different point which has been made to us is that the problems we have depicted are largely historical: that Broadmoor's non-psychotic population is the product of the admission practices of previous years; and that with more rigorous and careful selection procedures, difficulties with this group of patients need no longer occur. We regard this view as unrealistic. No doubt selection procedures could be improved, and the 1983 Mental Health Act now allows offenders to be sent to hospital for up to six months on interim orders, so that their suitability for treatment can be assessed. But 'even with the most assiduous investigation and preparation' (Mawson, 1983), problems about admitting non-ill offenders to indeterminate detention will remain. A man may be accepted in the belief that his offending behaviour is treatable. Subsequent developments, including what is learned about him (such as the nature of his sexual fantasies), may nevertheless years later prevent his discharge on grounds that have more to do with preventive detention than psychiatry.

Finally, the suggestion has been made to us that if only Broadmoor were able to treat its patients more intensively, then the dilemmas we have described, including the years without specific treatments, would be obviated. In our view it is not the lack of resources which constitutes the problem. Of course Broadmoor's resources could be increased; but in comparison with many NHS hospitals, they are already excellent (Hamilton, 1985). The medical complement and the nurse/patient ratio is high, and there is a large psychology department. Workshop and educational facilities are unusually good. Video equipment is provided for use in groups and social skills training. Psycho-physiological equipment is available for measuring patients' sexual responses. It is not, we believe, any lack of resources that caused in Broadmoor the problems we have described, but something much more fundamental – the concept of legal psychopathy, which brings offenders into indeterminate hospital detention yet lacks a hard medical basis. There is no value in intensifying treatment resources, unless there is understanding of what is to be treated, and how.

It is thus our view that the basic problems about the treatment and discharge of non-psychotic offenders which our research revealed cannot be attributed to deficiencies in Broadmoor's practice, although of course there were deficiencies. They arose from a deeper cause – the conceptual framework in which the courts are empowered to hospitalize such defendants. For compulsory medical treatment to be appropriate, there needs to be a medical formulation of the patient's problem and a medical criterion measure of the purpose of his treatment. If these are absent, the efficacy of treatment and the patient's readiness for discharge cannot be objectively assessed by doctors.

NON-PSYCHOTIC OFFENDERS IN THE SPECIAL HOSPITALS: PROPOSALS FOR CHANGE

There is no doubt that Broadmoor offers its non-psychotic patients a variety of therapies that many of them found helpful (Chapter 7): psychotherapy, social skills, and so forth. However, none of these treatments requires the subjects to be hospital in-patients – indeed, with the help of psychiatrists and psychologists, all of the treatments can be, and have been, carried out in prisons. (Gunn *et al.*, 1978; Perkins, 1982).

In our view, compulsory psychiatric detention can be justified only if the patient is suffering from a condition which requires him to be in hospital under the care of doctors and nurses. Non-psychotic offenders rarely suffer from such disorders and, as has been seen, our research showed that in deciding about the length of their detention, the doctors were unable to rely upon medical expertise: discharge decisions were closely related to non-medical factors such as the seriousness of the offence. If length of detention is to depend on factors like these, rather than on clinical considerations, then it would be better, we believe, for judges in open court rather than for doctors to determine it. We would therefore want to do away with hospital orders on the grounds of psychopathic disorder. We would not, however, want to do away with the opportunity of enabling these offenders to spend time in hospital – it is the legal framework within which they go there that we would like to change.

We believe that offenders who are neither mentally ill or impaired should not be detained in psychiatric hospitals against their wishes. We would therefore propose that they should only go to special hospitals as voluntary patients. This would presumably require legal changes, as under the National Health Service Act 1977 the special hospitals are provided 'for patients subject to detention' (see Chapter 1).

We would also want to change the legal provisions under which the offender who is not mentally ill is admitted to hospital. We would propose that if he is willing, and if the doctors and hospital authorities recommend it, then the courts should be enabled to send him straight to hospital with the status of a transferred prisoner, and with a length of custodial sentence determined in accordance with the ordinary principles of criminal justice. In addition, it should remain possible, as is now the case, for serving prisoners who are not mentally ill to be transferred from prison to hospital. But, in this case too, we believe that admission should take place only if the prisoner wants to come.

If changes along these lines were implemented, offenders who were not mentally ill or impaired would be admitted to, and remain in, hospital only on a voluntary basis. This would be the case whether the offender went to hospital straight from court, or whether he was transferred there whilst serving a prison sentence. In either case, the maximum length of his detention in hospital would be determined by the length of the custodial term which the sentencing court had imposed; doctors would not be able to detain these people beyond their normal prison release dates. The offenders themselves would be able to opt to leave the hospital and serve their sentences in prison if they so wished; and the hospital authorities could if they thought it appropriate seek a patient's removal to prison before his sentence expired.

Although they are basically different, our proposals have some features in common with those made in a consultative document 'Offenders Suffering from Psychopathic Disorder' issued in 1986 by the DHSS and Home Office (DHSS, Home Office, 1986). This document requires a brief mention here, giving as it did official recognition to the dubiety of the legal concept of psychopathic disorder and to some of the consequences of making it the touchstone of psychiatric detention. The background was concern in government circles about the possible release from the special hospitals of psychopathic patients who were liable to reoffend. In particular, there was a fear that Mental Health Review Tribunals, applying the statutory criteria of the 1983 Act, might discharge restricted patients on the grounds that they were no longer suffering from psychopathic disorder, even if the offenders concerned were considered by the Home Office still to represent a danger to the public.

The consultative document drew attention (para 15) to 'the inherent problems flowing from the uncertainties regarding the concept of psychopathic disorder, its diagnosis, treatability, and relationship with offending.' It also pinpointed the practical difficulties of assessing, particularly in an artificial hospital environment, whether the elusive disorder had in fact ameliorated.

Given these doubts and uncertainties, the authors of the document believed (para 21) that there was bound to be concern that the current arrangements for releasing these offenders gave insufficient weight to public safety. They therefore proposed that hospital orders on the grounds of psychopathic disorder should be abolished, and that such offenders should only be hospitalized as transferred prisoners, whom tribunals cannot release during the currency of the sentence. However, the transferred prisoners were to remain subject to indeterminate unrestricted detention on the grounds of psychopathic disorder once their sentences expired.

In the event this proposal was not proceeded with (Parliamentary Debates, 1986). Our own objection to it was that it paid scant regard to questions of justice for the individual. A concept that is discarded as too weak and uncertain to provide the grounds for release should also be recognized as being too weak to provide the grounds for detention. If the concept is as dubious as the consultative document shows it to be, then it provides an insufficient basis not only for keeping prisoners in hospital after their sentences have expired, but for bringing people into compulsory psychiatric detention at all.

Under our proposals, offenders who are not psychotic or mentally impaired would enter hospital only on a voluntary basis. This would enable them to receive help there without forfeiting their right to have the maximum length of their detention determined by the courts in accordance with the principles of criminal justice. They could thus benefit from the positive things hospital has to offer, without suffering the negative ones that stem from indeterminate detention for an uncertain disorder.

It should perhaps be made clear at this point that we are not proposing voluntary admission to hospital on the grounds that treatment there would enable these offenders to be 'cured' of their propensities to commit crimes. It seems apparent from the available reconviction studies (Bowden, 1981; Tennent & Way, 1984; Black, 1982) that residence in a special hospital has no discernible effect on the general law of reconviction – that the factor most strongly associated with reoffending is the amount of previous offending. In proposing a system of voluntary admission, we are simply recognizing that the special hospitals exist, and that unlike most other hospitals they are prepared to accept people who are neither mentally impaired nor ill. Our study showed that most PD offenders in Broadmoor felt that they had benefited in some way by being there. Our proposal is designed to ensure that offenders who might benefit from hospital facilities should be able to do so, but only so long as they themselves want to, and not at the cost of liability to indefinite psychiatric detention.

Not only the patients, but also the doctors would in our view benefit from the changes we propose. They would be relieved of the burden of having to recommend when non-psychotic offenders should be discharged; from the difficulty of having to negotiate their release with the Home Office; and from the great anxiety that can follow the release of such men – the notorious examples of reoffending by discharged special hospital patients have generally involved non-psychotic men (Chiswick *et al.*, 1984). It seems to us inappropriate and unfair that consultants should carry responsibility for making what are essentially non-medical judgments about the detention of serious offenders. The courts, and where appropriate the Parole Board, are where such judgments ought to be made.

An important aspect of our proposals is their bearing on the power of doctors to detain transferred prisoners who are not mentally ill, once their prison sentences have expired. At present, the Home Secretary's transfer direction has the same effect as a hospital order. The prisoner therefore becomes liable to indeterminate detention as soon as he is transferred, and when his release date arrives doctors can arrange for his continued detention on a 'notional' hospital order. Since we believe that the legal concept of psychopathic disorder has insufficient validity to justify the compulsory admission to hospital of offenders, we would want to abolish not only the court-imposed order, but also the administratively imposed one. If offenders cannot be detained on the grounds of mental illness or impairment, they should not, in our view, be kept in hospital against their wishes, either during or after their sentences.

A couple of riders have to be added to our proposals. First, it is important for a systematic and effective screening system to exist within the prisons (including remand prisons) to ensure that offenders who might benefit from treatment are suggested for it. Secondly, it should be made clear that we are not proposing that hospital admissions for reasons of asylum should cease. We believe that our proposals would enhance and intensify the role of active treatment, in that those people who did not benefit from it would not have to stay in hospital if the doctors thought it inappropriate (Mawson, 1983). But there will always be people who inspire little therapeutic optimism, yet who for various reasons are more appropriately placed in hospital than in prison. In Broadmoor hospital such men are offered humane asylum, and provided they are willing to come, we would hope that they will continue to be accepted.

An objection that has been made to our proposals is that if non-psychotic offenders came to hospital not on restriction orders, but with the status of transferred prisoners, often serving determinate sentences, then no supervision or recall provisions would be available after their release. However, as we have seen, the great majority of the offenders who are in Broadmoor under the legal category of psychopathic disorder have committed offences that rendered them liable to life imprisonment. This does not mean that if hospital orders were abolished, the courts would necessarily want to give life sentences to all or even most of these men. But when judges considered post-release supervision to be essential, they would, in most cases, be able to impose life sentences. Such sentences have been ruled by the Court of Appeal to be proper for 'persons who have committed offences of substantial gravity and who appear to be suffering from some disorder of personality or instability of character which makes them likely to commit grave offences in the future, if left at large or released from a fixed term of imprisonment' (Thomas, 1979). Offenders of this kind who were not mentally ill and who were under our proposals given life sentences could still go to the special hospitals. Their eventual release would then be subject to Home Office life licence conditions and recall, much as is the release of restricted patients. The safeguards offered under the two systems are very similar.

An additional point about life-sentence prisoners is worth making. The release of those who are transferred to hospital under the Mental

Health Act is normally governed by the same arrangements as apply to the release of those who remain in prison (Parliamentary Debates, 1985). This means that the Home Secretary can order release only on the recommendation of the Parole Board; and that he will not normally agree to release prisoners, whether from prison or hospital, until they have been detained long enough to satisfy what the judiciary considers to be the requirements of retribution and deterrence (Parliamentary Debates 1983, 1987). An unsatisfactory feature of the system is that this 'retribution and deterrence' period is determined in secret and cannot be appealed against. It would seem to us fairer if judges, when imposing a life sentence, could make it known in open court what length of detention the offence merited, and if that minimum sentence then became subject to appeal.

Another objection to our proposals which has been raised by Broadmoor consultants and others, is that it is desirable to make non-psychotic offenders subject to indeterminate detention because it compels them to realize that the only way they can get out is for them to 'change'. The indeterminate order, the argument goes, forces people to become motivated to have treatment. Perhaps this is sometimes the case, but there are different ways of looking at the process. As one man put it: 'You have to bow down to get through and out. The system wants you to agree with everything the doctor says about treatment. . . . I want to get out of Broadmoor, so I agree . . . I play the game'. But not all men are prepared or able to play it. One of the patients in our sample had been in Broadmoor for six years when his consultant reported: 'I am in despair about him. Long-continued efforts have failed to get him into any sort of mean-ingful treatment.' Perhaps this patient would eventually show himself 'responsive', perhaps not. Apart from the criminal justice considerations, it seems to us an unjustifiable use of valuable hospital beds, to allocate them in the hope that offenders will even-tually see the need for treatment, for if the hope proves to be un-justified there is no alternative to indeterminate detention.

Finally, it should be said that the number of offenders sent to the special hospitals under the legal category of psychopathic disorder nowadays is not large. In 1985, 10 such men were admitted on hospital orders and 16 as transferred prisoners. It might therefore be argued that since so few people are affected, it is hardly worth chang-ing the system. We would not take that view. The numbers admitted may not be large, but the consequences are profound: in Broadmoor, the average length of stay was eight years. Every admission thus has major long-term implications, both for the hospital and for the individual.

Appendix I

From mental illness to personality disorder: Nine case histories

In this appendix, we present details of the nine men in our study who had been admitted to Broadmoor with a legal classification of mental illness but were diagnosed as personality disordered by their Broadmoor doctors.

The first of these had been admitted to Broadmoor in his early twenties and had spent some twenty years there when we interviewed him. The forms completed at the time of his trial stated that he was suffering from schizophrenia, but the evidence was loose and rather uncertain. No reference was made to the presence of any expressive thought disorder or delusional symptoms and only the following details were presented as evidence for the diagnosis '[he revealed] . . . no specific paranoid symptoms or ideas but a confused uncertainty as to any sexual feelings' his 'emotional tone' was described as being 'trivial and flat'. Twenty years of observation in Broadmoor have revealed no clear signs of schizophrenia and although he has been subject to periods of depression and has entertained suicidal ideas, he is regarded by the Broadmoor doctors as someone who has a personality disorder and a particular problem regarding his sexuality. His rather rigid obsessional personality, in combination with his sexual drive, are considered to make him potentially dangerous to the object of his fantasy life. At the time of his trial the formulation of a possible incipient schizophrenic illness would have been perfectly reasonable clinically but the law makes no provision for question marks in relation to diagnosis, so he was described for official purposes as definitely suffering from a mental illness, in this case, schizophrenia.

The second man had also been in his twenties when admitted to Broadmoor. The psychiatric reports on him completed at that time reflected a wide degree of disagreement between the doctors who interviewed him. One doctor stated that the man was schizophrenic, one said he was 'probably schizophrenic', and the third said that he did not think he was schizophrenic. Despite this diversity of opinion it was stated on both section 60 order forms that the man was suffering from schizophrenia. Throughout his 12 years in Broadmoor he has given no evidence that he suffers from a psychotic illness and he himself maintains that he deliberately faked symptoms when on remand because he believed he would receive a lighter sentence if he were regarded as 'mental'. His previous psychiatric contact had been as an out-patient and the psychiatrist who had treated him in that setting had described him as an inadequate man with particular sexual difficulties. One of his trial psychiatrists stated that 'Whether this is a schizophrenic illness or not . . . I think there is no question that he is psychiatrically disturbed'. The lack of any evidence of mental illness during his long stay in Broadmoor gives considerable support to the man's own account of having faked symptoms at the time of admission. His offence on admission had been the killing of a stranger.

The third man in the series was in his early 30s when he was admitted to Broadmoor and had been a patient there for more than ten years when we interviewed him. At the time of his trial he was described thus: 'beneath a bland exterior he is very tense and anxious: sometimes he covers this with self-conscious giggling which gives a bizarre impression. He speaks about a terrible urge to hurt somebody.' The diagnostic statement made by the psychiatrist in his report was that he was a man who was suffering from' a well-known psychiatric disorder known as obsessional neurosis'. However, the same psychiatrist, when completing the official hospital order forms for the court, wrote 'in my opinion he is suffering from schizophrenia'. Psychiatrists will appreciate the diagnostic subtleties involved. The doctor used the more 'powerful' diagnosis on the legal forms while expressing doubt about the differential diagnosis in his report. Indeed, he stated in this latter document that the man might become schizophrenic, a statement hardly compatible with the opinion that he was already schizophrenic. The behaviour of the man in Broadmoor revealed no evidence of a psychotic illness and the authorities there regarded him as having engineered his own admission and very much regret having accepted him. They find it virtually impossible to discharge him because he says he will commit some life-threatening offence if he is released.

The fourth man in the series was in an anxious and depressed state at the time of his remand. Diagnosis was complicated by the fact that he was not British-born and cultural differences had to be taken into account when interpreting his offence. His behaviour in Broadmoor has not been characterized by abnormal levels of anxiety or any abnormal variation in mood. He is viewed instead as having a particular sexual problem and as possessing personality characteristics, such as impulsiveness, which in combination with his sexuality lead him to offend when under stress.

The fifth man was admitted from prison towards the end of his sentence in order to prevent his discharge into the community. Other than his denial of his offence and his often stated belief that he had been conspired against by various authorities at the time of his trial, he has displayed no evidence of psychiatric abnormality. However, these factors were enough to effect his transfer under the category of Mental Illness. His official transfer form stated 'these beliefs are of a delusional nature and based, in my opinion, on a paranoid psychotic illness'. Throughout his many years in prison he received no medical treatment, and he has received no treatment of any kind during his years in Broadmoor where, apart from continuing to deny his offence, he has provided no evidence of abnormality of mind.

This man was believed to be dangerous by the prison medical authorities on the good evidence of his past behaviour. His admission to Broadmoor might have been effected under the legal classification of psychopathic disorder rather than mental illness. In interview, his Broadmoor psychiatrist stated that he now regarded him as being personality disordered rather than as mentally ill, the nature of his disorder being his paranoid attitude toward his offence.

The sixth man in the group was regarded at the time of his trial as someone who might develop a schizophrenic illness later in life. He was in his early twenties at that time. To quote one of the psychiatrists who reported at his trial 'probably, in due course, the clinical picture will develop into a more typically schizophrenic one'. The diagnosis of schizophrenia was based on his abnormal jealousy, which was deemed to be of delusional intensity, and his obsessionality. In fact, after some six years in Broadmoor, no 'typically schizophrenic' picture has emerged and the fact that he had killed someone has come to be viewed as a result of the interaction between the particular circumstances he found himself in and certain personality attributes in the man himself, i.e., his emotional immaturity and his sadistic sexuality. In addition, he is seen as being a rigid, somewhat obsessional person who is of limited intelligence.

The seventh man in the group was also in his early twenties when admitted to Broadmoor. He had been there for more than four years when we saw him. Both psychiatrists who reported on him at the time of his trial stated that he was suffering from schizophrenia. The man reported having experienced auditory hallucinations and having entertained delusional beliefs about the person he had killed. He had a psychiatric history and had attended an out-patient clinic for several years. He was regarded by the psychiatrist responsible for him there as 'an anxious, rather obsessional, immature boy but not schizophrenic'. Close observation of his behaviour in Broadmoor has confirmed this latter view of him and, according to the man himself, he made up and exaggerated a number of his symptoms at the time of his trial in the belief that a psychiatric 'explanation' of his offence would in some way benefit him. His current diagnosis is 'obsessive neurosis with depressive symptoms'.

Relevant details about the eighth man are presented in Chapter 2. The basis for his admission diagnosis of depression was very weak and seems to have resulted from a wish or need on the part of the doctors to account for behaviour which could not really be explained. He has shown no abnormality of mood during his four years in Broadmoor and in interview maintained that he was ill neither at the time of his offence nor trial and that he should have been sent to prison.

The last man in this group was transferred to Broadmoor from prison when in his teens. There was considerable disagreement among the doctors who saw him at that time. The two responsible for his transfer stated quite categorically that he was suffering from schizophrenia, whereas the third doctor, a member of the prison staff, believed that he was faking his symptoms. His behaviour since his admission to Broadmoor four years ago has tended to endorse the opinion of the prison doctor. The man has given no evidence of any major mental illness and is now regarded as being neurotic and obsessional. On admission there was considerable disagreement between the Broadmoor staff regarding his diagnosis but, as one of the doctors there pointed out, 'whether or not he is psychotic, he is severely personality disordered and is dangerous in view of his sadistic sexuality and impulsive personality'. In other words, regardless of his diagnosis, he was considered dangerous. The offence which had brought him to prison was relatively minor and was histrionic and self-damaging in character. His behaviour in Broadmoor has confirmed the view that he is an obsessional man with a strong sado-masochistic component to his sexuality. When questioned about his behaviour in prison before his transfer he said he had been exaggerating his symptoms in the belief that it would help him get out of prison. He was correct in his belief but the net effect of his action has been to ensure his loss of liberty for a much longer period than he would have endured had he remained in the penal system.

Appendix II

Table 1.1. *Male admissions to Broadmoor between 1972 and 1984: by legal category and type of detention order*

Mental illness	Civil order	Court order	Restricted court order	Prison transfer	Other	Total
			Type of Detention Order			
1972	10	9	44	14	15	92
1973	8	4	46	5	12	75
1974	4	4	48	6	7	69
1975	5	4	38	6	12	65
1976	7	4	43	1	6	61
1977	6	3	29	7	5	50
1978	5	1	36	7	2	51
1979	6	4	35	9	7	61
1980	5	3	11	3	4	21
1981	3	2	15	9	7	36
1982	7	1	31	9	5	53
1983	3	0	21	4	2	37
1984	3	1	9	4	3	20

Psychopathic disorder	Civil order	Court order	Restricted court order	Prison transfer	Other	Total
1972	—	1	21	6	—	28
1973	—	2	43	5	—	50
1974	—	—	26	2	—	28
1975	—	1	11	3	—	15
1976	—	—	15	2	—	17
1977	—	—	14	—	—	14
1978	—	—	20	1	—	21
1979	—	—	6	1	—	7
1980	—	—	9	1	—	10
1981	—	—	9	—	—	9
1982	—	—	9	1	—	10
1983	—	—	6	2	—	8
1984	—	—	3	1	—	4

Table 2.1. *Mental illness legal classification: admission offence and criminal history*

	Admission Offence		Any prior record of:*		N = 116
	N	%	N		%
Homicide	38	33	3		3
Other serious violence	41	35	10		9
Assaults	12	11	33		28
Rape	4	3	4		3
Other sexual offences	3	3	8		7
Arson	10	9	3		3
Acquisitive	8	7	44		38

	Yes	%	No	%	NK
Any previous conviction	71	61	45	39	
Any previous custodial sentence	35	30	80	70	1

	M	sd
Mean number of previous court appearances	3.2	4.5
Mean age at first conviction	25.5	11.1

	N	%
Distribution of age at first conviction		
up to 16	19	16
17 – 19	29	25
20 – 29	37	32
30+	31	27

*These figures do not relate to the admission offence column and are not mutually exclusive

Table 2.2. *Mental illness legal classification: psychiatric and demographic variables by diagnosis at trial*

	Total		Paranoid Schizophrenic		Other Schizophrenic		Other Diagnoses			
	(N = 116)		(N = 43)		(N = 60)		(N = 13)			
	Yes % No %		Yes % No %		Yes % No %		Yes % No %		χ^2	p
Any previous in-patient care	78 67 38 33		23 55 19 45		47 78 13 22		8 62 5 38		7.23	0.03
Any record of diagnosis of psychosis prior to admission	69 62 43 38		21 51 20 49		44 75 15 25		4 33 8 67		10.12	0.01
Single (never married)	80 69 36 31		25 58 18 42		50 83 10 17		5 38 8 62		13.79	0.002
Was employed week of offence	28 26 81 74		10 26 29 74		12 21 45 79		6 46 7 54		3.23	0.08*
Country of birth UK or Eire	93 80 23 20		30 69 13 31		53 88 7 12		10 77 3 23		5.53	0.09
	M sd		M sd		M sd		M sd		F	p
Age at first hospital admission	26.2 8.9		29.5 10.9		22.4 6.1		22.4 12.6		2.29	0.10

*Schizophrenic groups v. Group 3

Table 2.3. *Mental illness legal classification: criminal variables by diagnosis at trial*

	Total (N = 116)				Paranoid Schizophrenic (N = 43)				Other Schizophrenic (N = 60)				Other Diagnoses (N = 13)				χ^2	p
	Yes	%	No	%	Yes	%	No	%	Yes	%	No	%	Yes	%	No	%		
Any previous convictions	71	61	45	39	24	56	19	44	38	63	22	37	9	69	4	31	0.99	N.S.
Any previous custodial sentence	35	30	81	70	11	26	32	74	19	32	41	68	5	38	8	62	0.91	N.S.
Someone killed in admission offence	38	34	77	66	16	39	26	61	14	23	46	77	8	62	5	38	8.06	0.02
Victim of offence was female	46	39	70	61	12	28	31	72	26	43	34	57	8	62	5	38	5.42	0.08
There was a sexual assault (real or intended)	14	12	102	88	3	7	40	93	7	12	55	88	4	31	9	69	5.34	0.08
	M	sd			M	sd			M	sd			M	sd			F	p
Age at first conviction	25.5	10.2			28.6	11.6			21.5	7.3			33.2	15.6			10.26	0.0001

Table 2.4. *Mental illness legal classification: comparison between men sent to local* hospital and Broadmoor under restriction orders*

	Broadmoor Sample				Local Hospital Sample				χ^2	p
	Yes	%	No	%	Yes	%	No	%		
Single (Never married)	68	68	32	32	43	54	36	46	3.41	0.07
Country of birth UK or Eire	80	80	20	20	71	88	8	12	3.26	0.09
Any previous conviction	59	59	41	41	49	62	30	38	0.17	N.S.
Any previous custodial sentence	30	30	70	70	30	38	49	62	1.25	N.S.
Any previous hospital admission	69	69	31	31	48	62	30	38	1.08	N.S.
Admission offence										
Homicide	47	47			7	7				
Other serious violence	25	25			40	38				
Assault	9	9			14	13			43.9**	0.0001
Rape	2	2			0	0				
Other sex offences	2	2			10	10				
Arson	8	8			13	12				
Other	7	7			20	19				
Diagnosis										
Paranoid Schizophrenia	38	38			21	20				
Schizophrenia	52	52			45	43			28.5	0.0001
Affective disorder	8	8			30	28				
Other	2	2			9	9				

* The population comprises all men sent to local hospitals under restriction orders in 1962–64 and 1975.
** Homicide v. rest

Table 3.1. *Psychotic men: diagnosis at time of study*

Diagnosis at time of Study:	Group I Paranoid Schizophrenics	Group II Schizophrenics	Group III Other Psychoses	N = 127
Paranoid Schizophrenia	44			
Paranoid State	2			
Schizophrenia		65		
Manic Depression			6	
Dementia			2	
Organic Psychoses			2	
Other Psychoses			6	
Totals:	46 (36%)	65 (51%)	16 (13%)	

Table 3.2. *Psychotic men: differences between diagnostic groups*

	Paranoid Schizophrenics (N = 46)		Schizophrenics (N = 65)		Other Psychoses (N = 16)		N = 127	
	N	%	N	%	N	%	χ^2	p
Treatment								
On major tranquillizers	43	94	56	86	5	31	32.6	0.0001
On minor tranquillizers	1	2	3	5	3	19	6.5	0.04
On antidepressants	1	2	6	9	3	19	4.8	0.09
On lithium	0	0	3	5	2	13	5.1	0.08
On more than one type of major tranquillizer	14	30	31	48	2	13	8.5	0.01
On anti-Parkinson medication	36	80	46	71	4	25	16.9	0.001
On any type of psychotropic medication	44	96	58	89	11	69	8.8	0.01
Doctor's opinion of mental state								
Deluded	28	64	26	43	0	0	15.1	0.001
Hallucinated	7	16	18	30	1	7	5.1	0.08
Disturbed mood	12	27	23	39	9	56	4.7	0.09
Organic factor	2	4	4	6	6	38	16.8	0.001
	M	sd	M	sd	M	sd	F	p
Age at interview	43.5	10.7	37.6	9.2	46.2	15.0	6.6	0.002
Length of stay in Broadmoor	8.5	4.0	8.5	4.8	8.5	3.4	0.06	0.99

Table 3.3. *Psychotic men: quantity of major tranquillizers by diagnostic subgroup*

Daily Quantity In cpz equivalent units*:	Paranoid Schizophrenia (N = 46)		Other Schizophrenia (N = 64)		Other Psychoses (N = 16)		Total 126	
	N	%	N	%	N	%	N	%
None	3	7	9	14	11	69	23	18
Up to 300 mg	12	26	13	20	2	12	27	21
301–600 mg	16	35	12	19	0	0	28	22
601–900 mg	2	4	9	14	2	12	13	10
901–1200 mg	6	13	9	14	0	0	15	12
1201–1800 mg	5	11	6	9	0	0	11	9
1801+ mg (range to 4398)	2	4	6	9	1	6	9	7

* Conversion tables provided by Aschkenasy and Carr (1982)

Table 3.4. *Psychotic men treated by ECT: number of courses and treatments*

	Paranoid Schizophrenia (N = 19)		Other Schizophrenia (N = 37)		Other Psychoses (N = 6)		N = 62	
	M	sd	M	sd	M	sd	F	p
No. of ECT Courses	1.5	0.6	3.9	2.8	1.7	1.0	8.69	0.001
No. of ECT treatments	9.1	5.0	29.8	21.9	14.7	12.1	9.22	0.001
Length of stay in Broadmoor (years)	10.2	3.5	9.9	4.2	9.2	3.7	0.14	n.s.

Table 4.1. *Psychotic men not being discharged: doctors' conditions for release*

	Improvement in illness not mentioned as condition				Improvement in illness stated as condition					
	Yes	%	No	%	Yes	%	No	%	χ^2	p
Working during week of offence	5	15	29	85	17	33	35	67	3.49	0.06
Country of birth U.K. or Eire	35	97	1	3	43	77	13	23	7.09	0.01
Receiving a major tranquillizer	26	72	10	28	50	89	6	11	4.44	0.03
Persisting doubts with regard to diagnosis	14	39	22	61	5	9	51	91	12.00	0.002
Evidence of psychosis in Broadmoor	30	83	6	17	54	96	2	4	3.22	0.07
RMO states deluded at time of interview	17	47	19	53	43	80	11	20	10.21	0.01
Does patient know what his drugs are for?	22	79	5	18	26	55	21	45	5.15	0.03
Does patient think his drugs have helped him?	18	75	6	25	20	44	25	56	5.91	0.02
RMO says security needs could be met in MSU	14	40	21	60	33	60	22	40	3.42	0.07
Final diagnosis was:										
Paranoid schizophrenia	3	8			27	48			23.46	0.0001
Schizophrenia	18	50			25	45				
Other psychosis	15	42			4	7				
Degree of control by drugs rated as:										
None/adequate	11	20			43	78			22.72	0.0001
Good/very good	17	47			11	21				
Other (e.g. does not require drugs)	8	22			1	2				

Table 4.2. *Psychotic men: elements in doctors' discharge decisions*

	Total Score	As percentage of maximum possible score
Change in mental illness	55	68
Reduced risk of reoffending	44	54
Change in behaviour	41	51
Nurses' opinion	32	39
More easily supervised	31	38
Change in personality	21	26
New facility is available	20	25
Change in external circumstances	3	4

Table 4.3. *Psychotic men by discharge status: trends between groups*

	To be discharged		Could move to less security		Needs maximum security		F	p
	M	sd	M	sd	M	sd		
(a) MMPI								
F scale	8.5	5.8	11.1	7.4	13.3	6.7	3.07	0.051
Anxiety scale	12.5	8.3	15.6	9.6	18.9	9.1	2.93	0.059
Impulsivity	10.5	5.4	12.8	6.4	14.2	4.5	2.51	0.088
Paranoia	1.9	2.7	3.9	4.0	5.3	3.9	4.70	0.012

	Yes	%	No	%	Yes	%	No	%	Yes	%	No	%	χ^2	p
(b) Doctors' mental state assessment														
Deluded at time of interview*	7	27	19	73	35	63	21	37	25	72	10	28	13.4	0.002
Mental state much improved	10	36	17	64	5	9	51	91	5	14	29	86	10.4	0.01

* 'Yes' includes those men rated 'possibly deluded'

Table 4.4. *Psychotic men by discharge status: distinguishing variables for maximum security*

	To be discharged				Could move to less security				Needs maximum security				χ^2	p
	Yes	%	No	%	Yes	%	No	%	Yes	%	No	%		
Diagnosis at time of study:														
Paranoid schizophrenia	11	37			13	22			19	54			10.6	0.031
Schizophrenia	15	50			32	55			13	37				
Other psychosis	4	13			13	22			3	9				
Consultant has known man for only 1/2 years	15	50	15	50	22	39	34	61	25	73	9	27	9.9	0.01
Was he working during week of offence	5	18	23	82	8	15	46	85	13	39	20	61	7.5	0.02
Did patient know what drugs were for	19	73	7	27	38	76	12	24	12	46	14	54	7.4	0.03

Table 4.5. *Psychotic men by discharge status: distinguishing variables for discharge*

	To be discharged				Could move to less security				Needs maximum security				χ^2	p
	Yes	%	No	%	Yes	%	No	%	Yes	%	No	%		
Degree of drug control:														
Very good	9	38			5	9			1	3			21.7	0.001
Adequate/good	15	63			42	78			21	68				
Inadequate	0	0			7	13			9	29				
Is anyone pressing for the man to get out?	10	42	14	58	8	14	48	86	3	9	32	91	11.6	0.003
Has he ever had ECT in Broadmoor?	8	27	22	73	35	60	23	40	17	50	17	50	8.9	0.011
Link with family rated as 'some' or 'strong'	23	79	6	21	31	56	24	44	16	53	14	47	5.3	0.08
Patient says he feels ready to leave	24	90	3	10	33	61	21	39	16	72	10	28	7.1	0.05

Table 4.6. *Psychotic men by discharge status: distinguishing variables for less security*

	To be discharged				Could move to less security				Needs maximum security				χ^2	p
	Yes	%	No	%	Yes	%	No	%	Yes	%	No	%		
Patient was on more than one major tranquillizer	6	21	23	79	29	50	29	50	12	34	23	66	7.38	0.025
Patient was on anti-parkinson medication	19	63	11	37	46	81	11	19	20	57	15	43	6.45	0.039
Offence rated at top of violence scale	23	79	6	21	28	48	30	52	21	60	12	34	7.72	0.03
Victim was a stranger*	6	21	22	79	18	42	25	58	7	23	24	67	4.63	0.10

* $\chi^2 = 4.62$ 1 df p< 0.04

Table 6.1. *Psychopathic disorder legal classification: criminal characteristics*

	Admission offence		Any prior record of:*		N = 117
	N	%	N	%	
Homicide	35	30	3	3	
Other serious violence	30	26	17	15	
Assault	16	14	38	33	
Rape	7	6	4	4	
Other sexual offences	12	10	24	21	
Arson	11	9	0	0	
Acquisitive	6	5	75	64	
	Yes	%	No	%	
Any previous conviction	100	85	17	15	
Any previous custodial sentence	65	56	52	44	
	M	sd			
Mean number of previous court appearances	5.1	4.8			
Mean age at first conviction	17.8	6.1			
	N	%			
Distribution of age of first conviction					
up to 16	57	49			
17 – 19	34	29			
20 – 29	18	15			
30 +	8	7			

*These figures do not relate to the admission offence column and are not mutually exclusive

Table 6.2. *Psychopathic disorder legal classification: admission form statements about treatment*

	Yes	%	No	%	NK
Was susceptibility to treatment mentioned in hospital order form?	26	23	86	77	5
Were any of the following stated as reasons for treatment being appropriate?					
Youth of offender	11	10	102	90	4
Motivation	25	22	87	78	5
High IQ	3	3	107	97	4
Low IQ	3	3	107	97	4
Failure of previous prison/ borstal to influence behaviour	18	16	95	84	4
Lack of previous treatment opportunity	4	4	109	96	4
Were any of the following types of treatment specified as being appropriate?					
Psychotherapy	9	8	102	92	6
Behaviour therapy	0	0	111	100	6
Anti-libidinal medication	3	3	109	97	5
Structured milieu	11	10	98	90	8
Any other specific treatment	7	6	103	94	7

Table 6.3. *Psychopathic disorder legal classification: comparison of Broadmoor and local hospital patients*

	PD men admitted to local hospitals in 1963/4 under hospital orders				PD men admitted to local hospitals in 1963/4 under restricted hospital orders				Broadmoor population restricted patients only						
	Yes	%	No	%	Yes	%	No	%	Yes	%	No	%	NK	χ^2	p
Any previous crime	32	89	4	11	17	100	0	0	89	89	13	13		2.42	N.S.
Any previous sexual offence	2	5	34	95	5	33	10	67	23	22	79	78	2	6.86	0.05
Any previous violent offence	5	14	31	86	3	18	14	82	44	43	58	57		12.37	0.002
Any previous custodial sentence	12	33	24	67	9	53	8	47	59	58	43	42	1	6.41	0.05
If ever married	10	28	26	72	3	18	14	82	27	27	75	73		6.41	0.05
Any previous IP care	30	86	5	14	14	82	3	18	43	42	59	58		25.31	0.0001

Source of the local hospital samples: Robertson & Gibbens, unpublished report to the Home Office, 1979 (from an original study by Walker & McCabe 1973)

Table 7.1. *Non-psychotic men: consultants' diagnosis at time of study*

	N (106)	%
Neurotic	4	4
Paranoid personality disorder	1	1
Affective personality disorder	1	1
Schizoid personality disorder	2	2
Explosive personality disorder	2	2
Hysterical personality disorder	3	3
Inadequate personality disorder	10	9
Other personality disorder	3	3
Psychopathic personality or unspecified personality disorder (ICD–9: 301.9)	74	70
Sexual disorder (various)	4	4
Drug dependence	1	1
Epilepsy	1	1

Table 7.2. *Non-psychotic men: admission and previous offences by age group*

	Group I		Group II		Group III			
	Admission Age 21 or less		Admission Age 22 – 29		Admission Age 30 +			
	N = 43		N = 32		N = 31			
	N	%	N	%	N	%	χ^2	p
Admission Offence								
Homicide	16	37	11	34	7	23		
Other Violence	16	37	9	28	10	32		
Sexual	4	9	7	22	8	26	16.3	0.04
Arson	1	2	4	13	6	19		
Other	6	14	1	3	0	0		
	M	sd	M	sd	M	sd	F	p
Previous arson convictions	0.07	0.46	0.09	0.53	0.52	1.2	3.63	0.03
Previous convictions for indecent assault	0.16	0.62	0.56	1.5	1.6	2.7	5.38	0.006

Table 7.3. *Non-psychotic men: type of treatment by age at admission*

| | Group I (N = 43) | | Group II (N = 32) | | Group III (N = 31) | | Total (N = 106) | | | |
| | 21 or less | | 22 – 29 | | 30 + | | | | | |
	N	%	N	%	N	%	N	%	χ^2	p
Currently having:										
Individual psychotherapy	9	21	4	12	2	7	15	14	3.39	0.18
Group therapy	11	26	1	3	2	7	14	13	9.82	0.007
Other psychological treatments: (social skills, sex education, speech therapy, behaviour modification)	6	14	3	9	0	0	9	8	4.56	0.10
None of the above	18	42	25	78	28	90	71	67	21.70	0.001

	M	sd	M	sd	M	sd	M	sd	F	p
Psychotropic medication	3	7	7	22	5	16	15	14	3.49	0.18
Has he ever had:										
Individual psychotherapy	25	58	13	41	8	26	46	43	7.30	0.03
Group therapy	36	84	23	74	16	52	75	71	9.26	0.01
Other psychological treatments: (social skills, sex education, speech therapy, behaviour modification)	26	62	14	44	4	13	44	42	17.66	0.001
None of the above	3	7	3	9	11	36	17	16	12.38	0.002
	M	sd	M	sd	M	sd	M	sd	F	p
Length of stay in Broadmoor	7.4	4.6	7.9	4.4	9.2	4.4	8.2	4.5	1.44	0.27
Percentage of time spent in treatment	46	31	32	26	16	19	33	29	11.77	0.001

Table 7.4. *Non-psychotic men: consultants' views on treatment*

	Group I		Group II		Group III			
	Admission Age 21 or less		Admission Age 22 – 29		Admission Age 30 +			
	N = 32		N = 26		N = 27			
	N	%	N	%	N	%	χ^2	p
a) The role of psychotherapy								
No role to play	1	3	1	4	8	30		
Little role	1	3	1	4	5	19	29.4	0.001
Important role	10	31	14	54	12	44		
Very important	20	63	10	38	2	7		
b) Is psychiatric treatment needed now?								
No	0	0	0	0	4	13		
Probably Not	1	2	3	10	7	23		
Probably Yes	7	16	3	10	10	32	31.5	0.0001
Yes	23	54	13	43	3	10		
Supervision only*	12	28	11	37	7	23		

* Applied to men being considered for discharge

Table 7.5. *Non-psychotic men: consultants' views of relation of time and treatment to fitness for discharge*

	Group I		Group II		Group III		Total	
	Admission Age 21 or less		Admission Age 22 – 29		Admission Age 30 +			
	N = 25		N = 17		N = 21		N = 63	
	N	%	N	%	N	%	N	%
Mainly time	6	24	5	29	12	57	23	37
50 – 50	8	32	6	35	2	10	16	25
Mainly treatment	11	44	6	35	7	33	24	38

$\chi^2 = 7.31$ 4 df p < 0.12

Table 8.1. *Non-psychotic men: admission offence by discharge status*

Type of offence:	To be discharged		Could move to less security		Needs maximum security			
	N = 39		N = 39		N = 23	(N = 101)	(NK = 5)	
	N	%	N	%	N	%		
Homicide	7	18	12	31	14	61		
Other Violence	15	38	13	33	5	22		
Sex	8	21	8	21	3	13		
Arson	5	13	4	10	1	4		
Other	4	10	2	5	0	0		

Homicide v. All other offences $\chi^2 = 12.22$ 2 df $p < 0.002$.

Table 8.2. *Non-psychotic men: personality change cited by doctors and selection for discharge*

	N	%	N = 39
More self control, less impulsive, etc.	15	39	
Generally more mature (nothing more specific said)	14	36	
More sociable, confident, etc.	4	10	
Greater understanding of danger of drinking	2	5	
Becoming an old man	1	2	
None mentioned	3	8	

Table 9.1. *Mental illness and psychopathic disorder patients on restricted orders: criminological differences. All male admissions to Broadmoor 1972 to 1981*

	Mental Illness Group		Psychopathic Disorder Group			
	(N = 273)		(N = 133)			
	N	%	N	%	χ^2	p
Any juvenile offending	42	15	63	47	46.1	0.0001
Any previous conviction as an adult	157	58	101	76	13.1	0.001
Any previous conviction for violence	64	23	49	37	7.3	0.01
Any previous conviction for a sexual offence	19	7	33	25	23.9	0.0001
Any previous prison sentence	57	21	53	40	16.3	0.001
Broadmoor doctor thinks offence was sexually motivated	37	14	53	40	31.5	0.0001
	M	sd	M	sd	F	p
Age of first court appearance	27.9	12.8	17.5	6.1	79.7	0.0001

Source: DHSS Special Hospital Research Unit

Table 9.2. *Mental illness and psychopathic disorder patients on restriction orders: social and psychiatric differences. All male admissions to Broadmoor 1972–1981*

	Mental Illness Group		Psychopathic Disorder Group			
	N	% (Total)	N	% (Total)	χ^2	p
(Social)						
Maternal separation	69	27 (254)	56	47 (119)	13.5	0.001
Ever in a children's home	13	5 (259)	21	17 (124)	13.3	0.001
Born in the U.K. or Eire	197	75 (262)	120	95 (126)	21.6	0.0001
(Psychiatric)						
Any previous in-patient care	185	70 (266)	56	42 (132)	26.1	0.0001
	M	sd (N)	M	sd (N)	F	p
Age on admission to Broadmoor	34.4	11.7 (273)	25.1	7.5 (133)	68.9	0.0001
No. of previous psychiatric admissions	3.1	3.9 (266)	1.3	3.0 (132)	22.4	0.0001

Source: DHSS Special Hospital Research Unit

References

Aschkenasy, B. & Carr, A. (1982). *Personal communication.*

Baggaley, A. R. & Riedel, W. W. (1966). A diagnostic assembly of items based on Comrey's factor analysis. *J. Clin. Psychol.*, **22**, 306–8.

Black, D. A. (1982). A five year follow up of male patients discharged from Broadmoor. In *Abnormal offenders, delinquency and the criminal justice system* (eds. Gunn, J. & Farrington, D.). Wiley, Chichester.

Blackburn, R. (1975). An empirical classification of psychopathic personality. *Brit. J. Psychiat.*, **127**, 456–60.

Bowden, P. (1981). What happens to patients released from special hospitals? *Brit. J. Psychiat.*, **138**, 340–5.

Butler Committee Report (1975). See Home Office, DHSS (1975).

Carson, R. C. (1969). Interpretive manual to the MMPI. In *MMPI research developments and applications* (ed. Butcher, N. J.). McGraw Hill, New York.

Chiswick, D. (1982). *The special hospitals: A problem of clinical credibility.* Bull. Roy. Coll. Psychiat., **6**, 130–2.

——, McIsaac, M. W. & McClintock, F. H. (1984). *Prosecution of the mentally disturbed.* Aberdeen University Press, Aberdeen.

Cleckley, H. (1976). *The mask of sanity.* (5th ed.) Mosby, St. Louis.

Cohen, D. (1981). *Broadmoor.* Psychology News Press, London.

Crawford, D. A. (1978). A social skills programme. In *Special hospitals research report No. 14.* DHSS, London.

——, & Allen, J. V. (1979). In *Love and attraction: Proceedings of an International Conference* (eds. Cook, M. & Wilson, G.). Pergamon Press, London.

Criminal Procedure (Insanity) Act (1964). HMSO, London.

Dell, S. (1980). The transfer of special hospital patients to NHS Hospitals. *Special Hospitals Research Report No. 16.* DHSS, London.

—— (1984). *Murder into manslaughter.* Oxford University Press, Oxford.

——, Robertson, G. & Parker, E. (1987). Detention in Broadmoor: factors in length of stay. *Brit. J. Psychiat.*, **150**, 824–7.

DHSS (1981). *Mental health enquiry statistics for England and Wales.* - DHSS, London.

—— (1983). *Memorandum on the Mental Health Act 1983.* DHSS, London.

—— (1984). *Special hospital patient statistics.* DHSS, London.

——, Home Office (1986). *Consultation document: Offenders suffering from psychopathic disorder.* DHSS, London.

Edwards, S. & Kumar, V. (1984). A survey of prescribing of psychotropic drugs. *Brit. J. Psychiat.,* **145,** 502–7.

Foulds, G. A. (1965). *Personality and personal illness.* Tavistock Press, London.

——, & Caine, T. M. (1967). *Manual of the hostility and direction of hostility questionnaire.* University of London Press, London.

Gibbens, T. C. N. & Robertson, G. (1983). A survey of the criminal careers of hospital order patients. *Brit. J. Psychiat.,* **143,** 362–9.

——, Soothill, K. L. & Pope, P. J. (1977). *Medical remands in the criminal court.* Oxford University Press, Oxford.

Gostin, L. O. (1977). *A human condition,* Vol 2. Mind, London.

Grounds, A. T. (1987a). Talk to annual meeting of the Forensic section of the Royal College of Psychiatrists in Stratford, February 1987.

—— (1987b). The detention of 'psychopathic disorder' patients in special hospitals: critical issues. *Brit. J. Psychiat.,* **151,** 474–8.

—— (in press). The transfer of sentenced prisoners to Broadmoor hospital.

Gunn, J. & Robertson, G. (1976). Drawing a criminal profile. *Brit. J. Criminol.,* **16,** 156–60.

——, Robertson, G., Dell, S. & Way, C. (1978) *Psychiatric aspects of imprisonment.* Academic Press, London.

Hamilton, J. (1985) The special hospitals. In *Secure Provision,* (ed. Gostin, L.). Tavistock Publications, London.

Hare, R. D. & Schalling, D. (1978). *Psychopathic behaviour, approaches to research.* Wiley, Chichester.

Hathaway, S. R. & McKinley, J. C. (1943). *A Manual for the Minnesota Multiphasic Personality Inventory.* The Psychological Corporation, New York.

Henderson, D. (1939). *Psychopathic states.* Norton, New York.

Hill, D. & Watterson, D. (1942). EEG studies of psychopathic personalities. *J. Neurol. Psychiat.,* **5,** 47–65.

Home Office, DHSS (1975). *Report of the Committee on mentally abnormal offenders.* Cmnd 6244. HMSO, London.

Lewis, A. (1974). Psychopathic disorder: a most elusive category. *Psychol. Med.*, **4**, 133–40.

MacCulloch, M. (1982). The Health Department's management of special hospital patients. In *Dangerousness: psychiatric assessment and management* (eds. Hamilton, J. & Freeman, H.). R. Coll. Psychiat., London.

McKinley, J. C. & Hathaway, S. R. (1956). Scales 3 (Hysteria), 9 (Hypomania), and 4 (Psychopathic Deviate). In *Basic readings on the MMPI in psychology and medicine* (eds. Walsh, G. S. & Dahlstrom, W.G.). University of Minnesota Press, Minnesota.

Mawson, D. (1983). *Psychopaths in special hospitals.* Bull. R. Coll. Psychiat., **7**, 178–81.

Mental Deficiency Act (1913). HMSO, London.

Mental Deficiency Act (1927). HMSO, London.

Mental Health Act (1959). HMSO, London.

—— (1983). HMSO, London.

Ministry of Health (1960). *Memorandum on the Mental Health Act 1959.* HMSO, London.

National Health Service Act (1977). HMSO, London.

Parliamentary Debates (1983). *House of Commons Official Report.* (Hansard). 30.11.83. Cols 513–51.

—— (1985). *House of Commons Official Report.* (Hansard). 23.7.85. Cols 463–4.

—— (1986). *House of Commons Official Report.* (Hansard). 19.12.86. Col 735.

—— (1987). *House of Commons Official Report.* (Hansard). 23.7.87. Cols 346–7.

Partridge, R. (1953). *Broadmoor.* Chatto & Windus, London.

Perkins, D. (1982). The treatment of sex offenders. In *The prevention and control of offending* (ed. Feldman, P). Wiley, Chichester.

Prins, H. (1980). *Offenders, deviants or patients?.* Tavistock, London.

Reeve, A. (1983). *Notes from a waiting room.* Heretic Books, London.

Robertson, G. & Gibbens, T. C. N. (1979). *The criminal careers of mentally abnormal offenders.* Unpublished report to the Home Office, London.

Royal Commission (1957). *Royal commission on the law relating to mental illness and mental deficiency.* Cmnd 169. HMSO, London.

Shepherd, M. & Sartorius, N. (1974). Personality disorder and the International Classification of Diseases. *Psychol. Med.,* **4**, 141–6.

Taylor, P. J. & Gunn, J. (1984). Violence and psychosis I. *Brit. Med. J.*, **288**, 1945–9.

Tennent, G. & Way, C. (1984). The English special hospital—a 12–17 year follow up study. *Med. Sci. Law*, **24**, 81–91.

Thomas, D. A. (1979). *Principles of sentencing*. Heinemann, London.

Thompson, P. (1972). *Bound for Broadmoor*. Hodder & Stoughton, London.

Tidmarsh, D. (1981). Letter, *Bull. R. Coll. Psychiat.*, **5**, 211.

Walker, N. & McCabe, S. (1973). *Crime and insanity in England Vol. 2*. Edinburgh University Press, Edinburgh.

Welsh, G. (1956). MMPI profiles and factor scales A and B. *J. Clin. Psychol.*, **21**, 43–7.

Williams, G. (1978) *Textbook of criminal law*. Stevens, London.

World Health Organization (1978). *Mental disorders: glossary and guide*, Ninth Revision. WHO, Geneva.

Young, M. & Hall, P. (1983). The never ending sentence of J. W. In *Rough justice*, pp. 75–99. Ariel Books, London.

Index